ENVIRONMENTAL
Literacy

ENVIRONMENTAL *Literacy*

Everything You Need to Know About Saving Our Planet

H. Steven Dashefsky

RANDOM HOUSE

NEW YORK

For LINDSAY and KIM

GE 10
.D37
1993

Library of Congress Cataloging-in-Publication Data

Dashefsky, H. Steven.
    Environmental literacy: everything you need to know about saving our planet/H. Steven Dashefsky.—1st ed.
        p.   cm.
    ISBN 0-679-74774-5
    1. Environmental sciences—Dictionaries.   I. Title.
    GE10.D37   1993   363.7′003—dc20   92-56808

Design by Robert Bull Design

Manufactured in the United States of America on acid-free recyclable paper using partially recycled fibers

9 8 7 6 5 4 3 2

PREFACE

Our planet has been compared to a spaceship, and the analogy is especially poignant when we consider the occupants of both vessels. Both the astronauts and the inhabitants of spaceship earth must rely on the integrity of their ships for survival and make do with the resources on board. Most people would consider the astronauts at a much greater risk than those of us who remained behind. But who is really in a more precarious situation?

The astronauts know exactly what they can and cannot do while living in their environment. They know which knobs must be pushed to maintain their ship's temperature, where their food will come from, and how to dispose of their wastes. Those of us on earth, on the other hand, have little insight into what the consequences will be when we play with the controls. We are just now beginning to realize how using resources and disposing of our wastes impact our world. We didn't have the benefit of taking simulated flights or learning emergency landing procedures prior to takeoff. Spaceship earth didn't come with an instruction manual so we must write one as we go.

Creating these instructions will be a difficult task. Scientists must first agree on what needs to be done. Then we must support these new initiatives, incorporate them into our way of life, and convince politicians to pass legislation assuring a long-term commitment to preserve the integrity of our planet.

The main objective of this book is to help you make informed, educated decisions. Whether you are creating your own personal set of priorities, writing to elected officials urging legislation, or voting on specific laws, your judgements must be based upon fact. The decisions you make will become part of the instruction manual that will determine the fate of our ship.

Making environmentally sensitive decisions and becoming involved is difficult. As you read or watch the daily news, there is a barrage of environmental stories. There are newly proclaimed ecological crises requiring attention, new and pending local, state, and federal legislation to be debated, and elected officials to be voted in or out of office. If you decide to take action about a particular issue, it must be done quickly since most of these issues pass through the spotlight rapidly, with only a brief opportunity to interject our opinions.

Most of us don't have time to research a topic. Even people who want to learn about environmental issues and take some form of action are often thwarted in their attempts. Becoming and remaining environmentally literate is difficult for many reasons. Ecology, by definition, is the study of relationships. To comprehend a typical environmental topic you might have to search through biology, chemistry, geology, physics, sociology, engineering, agriculture, and law books to dig up all the facts. Even then, you are not seeing the big picture, just the pieces that must be put together and analyzed as a whole. This book is designed to address environmental topics, not individual disciplines. All the topics included are viewed from an ecological perspective with explanations written for the layperson.

There are other problems you may encounter while trying to remain informed. Many important terms or phrases you hear or read about are not found in a standard dictionary or encyclopedia. This occurs for two reasons. First, many of the phrases used in the environmental movement today are new and untested. It takes years of use for terms or phrases to become accepted in the English language. We don't have time to wait, since understanding this terminology is important now.

Many highly respected scientists and leaders give us as few as 40 years to change our ways before it's too late. Even if some of the terminology changes over the next few years and is never accepted as part of our language, the relevance of these terms makes them too important to ignore. This book includes entries that are considered important but won't be found in any other reference book.

Specificity is another reason why many environmental terms and phrases are difficult to research. Many of these topics are so narrowly focused that they would be inappropriate in any other reference piece. The same is true of the many people, organizations, and legislative acts and laws that address these issues. All of these entries are pertinent to anyone interested in the environment and are included in this book.

Ecological problems are complex by nature and usually controversial. The only way most people, even those with an innate interest, will remain aware and up-to-date is if the information is quickly accessible and easily understood. This desktop reference is designed to do just that.

This book translates the facts into plain English, so you can make your own decisions on how to manage our environment. The information is presented in a clear, concise, and easy-to-read manner, and is written for someone with little or no formal education in the environmental sciences. Each entry is designed to first give background information, if needed, and then specific information about the topic from an environmental perspective.

HOW TO USE THIS BOOK

Imagine yourself cutting out the pages from one of the many environmental books on the market, rearranging them, and pasting them back into the book in the sequence you prefer. You threw out those pages that didn't interest you and put those that did up front. Of course, you wouldn't want to do this and don't have to.

Everyone has their own interests and priorities, and it's difficult to find a book tailored to meet your needs. *Environmental Literacy* offers an alternative way to read about our environment. It is designed to flow with your interests, whatever they might be, opening new doors each step of the way. Reading this book should be like a free-form learning experience with each topic an open invitation to the next.

The best way to begin reading this book is by flipping through the pages. When an entry captures your interest, read it, and notice the boldface words, which are themselves entries. These words will lead to new topics that are of interest, to let you delve deeper into the topic or to provide background material for a better understanding. Additional related entries are also listed. Let each entry guide you to the next, creating your own custom, on-demand reading and learning experience.

This book can also be used for reference and is a one-stop source for environmental terms and topics. The List of Entries (page 287) can help you find a particular term or phrase. Or, if you prefer more direction, see A Special Note to Educators on page 275.

A NOTE TO THE READER

Many names or phrases are best known by their acronym. For example, Dichloro-diphenyl-trichloro-ethane is best known as DDT. Another is chlorofluorocarbons, which are best known as CFCs. Words such as these are listed under their acronyms.

Boldface words, indicating they themselves are entries, may not be exact. For example, **pesticide residue** within an entry refers you to the entry **pesticide residues on food.**

ACKNOWLEDGMENTS

I want to thank the 600 + environmental organizations for their literature and valuable expertise, and my wife for her countless hours of help preparing the manuscript, and for support throughout this project.

INTRODUCTION

It's been 30 years since **Rachel Carson** wrote *Silent Spring,* which became the seed that would slowly grow into a national action movement. Millions of people jumped on the environmental bandwagon over the next few years, culminating in the first **Earth Day** in 1970. People felt moved and motivated as hundreds of organizations sprang up supporting environmental issues. A barrage of books was published with topics from environmental ethics and ecotactics to how to build geodesic domes, and the then-proposed supersonic transport. People monitored the level of **lead** in gasoline, **DDT** in **pesticides,** and compared the world's population with that of a time bomb waiting to go off. There was great concern about whether global temperatures were going to gradually rise due to the **greenhouse effect** or get colder due to pollutants in the upper atmosphere, causing another ice age.

It was as if everyone had found **ecology.** But then a funny thing happened. The fad began to fade. Interest in the environment waned and then slowly moved into the background. The underlying interest never left, but the activism was gone. Legislation was passed to relax many of the laws designed to protect our environment.

It wasn't until 20 years later that, all of a sudden, ecology was back! Where had it gone for 20 years? We had new threats of **global warming, ozone depletion, acid rain,** and another Earth Day in 1990. We now have ecosacs, ecopacs, and **deep ecology. Alternative energy** sources and environmentally appropriate architecture have once again caught our interests.

There is a serious problem afoot here. We like to get excited about things when there is a bandwagon to catch a ride on. Fads can be fun, and there is nothing wrong with participating in the festivities. The problem, however, is that man's impact upon our planet is not a passing fad. Our **biosphere** continues a slow, arduous, but steady deterioration. The environmental problems that existed 20 years ago still exist today.

Unfortunately, mother earth isn't into fads or 20-year cycles. Our interest in cleaning up the environment seems to peak every few decades, but we continue to pollute and otherwise damage our fragile **ecosystems** throughout the decades without any loss of interest. These problems must be addressed today, tomorrow, and as long as it takes to resolve them, and they must be addressed by everyone.

Some people want to preserve our environment for purely intrinsic reasons, even if there is no threat to our way of life. But we don't have to worry about deciding if intrinsic values are reason enough. The changes we are making to our planet will threaten our existence if left unchecked. You don't have to be an environmentalist to want to save our planet—you just need a will to survive.

A MATTER OF RESOURCES

There are many reasons why we have reached the brink of an environmental crisis. The first and foremost is a matter of resources—man's resources.

Resources are what one particular organism needs to survive as an individual and as a population. Each form of life requires its own array of resources, each with its own unique perspective. What is one organism's resources may be another organism's waste. A squirrel needs certain kinds of nuts for food, trees for shelter, and twigs and leaves to build its nest. The squirrel's waste may be food for another organism.

Each species is prevented from making excessive demands on these resources due to a complex set of natural checks and balances. Depending upon your level of technical expertise, these checks and balances can be called "mother nature," **food webs** with feedback mechanisms, or homeostasis. They can all mean the same thing—all species are kept in check, and a balance is maintained.

A sudden abundance of a certain type of plant results in a dramatic increase in the population of an insect that feeds upon that plant. The increase in the insect population quickly drives back down the plant population. The resulting decline of the plants, re-establishes the original population of insects. This may sound overly simplistic, but it demonstrates the balance of nature beautifully.

A few thousand years ago, man was part of the system, using a limited number of resources to sustain life and following the same set of checks and balances as all other forms of life. But when man began to develop sophisticated technologies, he no longer was part of the system but began to control the system. We now consider the entire biosphere our resource.

THE ULTIMATE CONSUMER

We have become the ultimate **consumers** and have made this fact the focal point of our society. We have consumer movements, consumer unions, and consumerism. Americans are some of the worst offenders, consuming about 30 percent of the world's resources while making up only about 6 percent of the world's population.

The term "consumer" exemplifies better than any other

how we have separated ourselves from nature. If you remember back to your high-school biology course, a consumer is an organism that eats its food instead of producing it. Consumers get their food by eating **producers,** which are the green plants that use the energy from the sun to produce food. Primary consumers get their food directly by eating the plants, while secondary consumers feed on the animals that eat the plants.

All consumers, with the exception of man, fill a small **niche;** fit neatly into intricate food webs; use a limited number of resources; and observe the natural checks and balances that keep any one species from overly zealous attempts to exploit the land.

Only man has the power to consume our natural resources unchecked. He **clear-cuts** the forests, **strip-mines** the land, and harvests the oceans with **driftnets.** He can clear thousands of acres of land to grow a single crop. Even more impressive than creating this unstable environment is his ability to enforce it. When natural catastrophes, such as the eruption of Mt. Saint Helens, devastate an area, nature slowly returns the area back to a balanced state. But when we clear the land to grow our crops, we then use **pesticides, herbicides,** and **fertilizers** to maintain this unnatural and unstable environment.

THE DANGER FROM WITHIN

We have been able to overpower the system for a long time with little apparent impact. Recently, however, it has become obvious there are serious pitfalls to this control that can become disastrous for all forms of life, including man. We can place these potential dangers into two categories—the exploitation of resources and the resulting **pollution** caused by that exploitation. Both don't necessarily have to destroy our planet if the damage is kept within certain guidelines. Nature has an amazing resiliency and ability to balance deviations that occur.

But we don't follow any guidelines. The quantity of pollutants we spew forth into the air, water, and soil is staggering, and the speed with which we pollute and exploit causes natural systems to fail. Nature works slowly, and changes are stabilized slowly. For example, the transition of an area from a simple ecosystem to a complex climax ecosystem, called **succession,** takes hundreds of years to accomplish. Man, however, changes entire ecosystems in a matter of days and dumps tons of waste in a few minutes.

We exploit our resources, enforce this exploitation, and pollute our biosphere, yet many people do not perceive an environmental crisis. Why? Man is accustomed to reacting quickly to change, but nature responds slowly—too slowly for man to take notice. We have great difficulty finding any interest in an event that doesn't conclude within our brief attention span. And this is the primary reason why the environmental problems of a few decades ago have become the crises of today. The damage we are inflicting on earth is occurring at a rapid pace, but earth's response is slow. To observe the changes would be similar to watching a plant grow. We have too short an attention span to do so and become concerned only when the plant begins to die. If you were observant, you could see obvious symptoms of a problem appear long before the demise of the plant.

THE TECHNOLOGICAL ANSWER

There are many people who believe that technology is the answer to our problems, and in many ways they are correct.

New and improved technologies have allowed mankind to perform feats that only a few decades ago would have been considered extraordinary. Whenever someone threatens impending environmental doom, technology seems to come to our rescue. A world food shortage is countered with agricultural technologies such as pesticides and pest **resistant crops,** sophisticated harvesting equipment, and **irrigation** systems that have allowed us to dramatically increase our food resources and to do so on smaller parcels of land. The world's grain harvest has increased almost 2.5-fold since 1950, and we can produce a five-pound chicken on less than nine pounds of feed. Technology has affected every part of our life and society. Modern medicine can improve the quality and increase the longevity of our lives.

Our successes, however, should not lull us into complacency about our problems. It is naive to believe that technology is a panacea for all our woes and doesn't come with its share of problems and shortcomings. It is the double-edged sword that can destroy as well as save us. Technology can exploit our resources on land and in the seas, oftentimes with devastating impact. It can cause a plethora of problems from **toxic wastes** in our air, water, and soil, to **landfills** overflowing with nonbiodegradable products and toxic substances. At its worst, it can cause acute and chronic illness and even death, as with some pesticides and manufactured **hazardous wastes.**

We may be too optimistic about technology always saving us just in time—like the cavalry coming over the hill. With the world's population increasing by over 80 million people each year and expected to exceed six billion by the year 2000, the stresses we place on our planet will demand technology to produce results at a frightening, if not unrealistic, pace.

Another disturbing fact about technology is our failure to implement useful technologies when they are available to us. There are many existing technologies that, if widely used, would dramatically improve many of our environmental problems today. Much less ***fossil fuel*** need be used to heat homes if we would use the same ***super-insulation*** technology found in parts of Canada, or develop ***solar power*** facilities, as found in Japan and Israel. We could use a fraction of the pesticides poisoning our planet if tested and proven effective ***integrated pest management*** techniques were widely used. These and many other technologies are not in widespread use because we don't want to change our way of life until the problems become severe. We continually need be reminded that it's far better to address a problem before it becomes a crisis.

Simply throwing technology at a problem won't necessarily resolve it. The same intelligence that went into creating the technology must be used to implement it, complement it with nontechnological solutions, and accept what shortcomings it may present, so alternatives can be found.

ENVIRONMENTAL
Literacy

abiota An **ecosystem** consists of two major components interacting with one another. The biota is the living component (plants and animals), and the abiota is the nonliving component, which includes the soil, water, and air. (See *Biosphere*)

abyssal ecosystem Refers to an **ecosystem** that exists deep within the ocean where no light penetrates. Organisms at this depth depend on organic debris (dead plants and animals) that drifts down from higher levels where light is visible. (See *Marine ecosystems*)

acid deposition See *Acid rain*.

acid mine drainage During **strip mining,** sulfur within the coal may become exposed to the elements and form (with the help of bacteria) sulfuric acid. This sulfuric acid can seep into streams or lakes and damage **aquatic ecosystems.**

acid rain When **fossil fuels** such as **coal,** oil, and natural gas are burned, many substances are emitted into the air. Sulfur dioxide, nitrogen compounds, and particulates are three such substances, and are considered primary pollutants responsible in part for **air pollution.** These substances travel through the air and react with each other in the presence of sunlight to form **secondary pollutants,** such as sulfuric and nitric acids. When these acids fall to earth with rain, it is called acid rain. Since these acids also come to the earth's surface in the form of snow, fog, dew, or small droplets the phrase "acid deposition" is often used.

Since these secondary pollutants float and are carried by winds, acid deposition often occurs far from its source. For example, the northeastern U.S. has some of the highest concentrations of acid rain, but much of it is produced by power and industrial plants in the midwest and carried by the westerly currents, eastward.

Normal rain is slightly acidic, having a pH of about 5.6. Average rainfall in most of New England and adjoining parts of Canada is between 4.0 and 4.5, which is about the acidity of grapefruit juice. Mountain tops in New Hampshire have recorded rains with a pH of 2.1, about the same acidity as lemon juice.

The most apparent damage caused by acid deposition is the destruction of statues that crumble from the acids, but the most serious effects are less noticeable. Studies show acid deposition at levels below 5.1 kill fish and destroy **aquatic ecosystems** since most organisms have narrow pH **tolerance ranges.** About 25,000 lakes in North America have been damaged by acid deposition.

Acid deposition weakens and kills trees and stunts the growth of crops and other plants. Although it is difficult to confirm, many scientists believe large portions of forests in northeastern North America and parts of central Europe are dying due to acid deposition. It also contributes to respiratory illness and, according to some professionals, is a major cause of lung disease in the U.S.

active solar-heating systems   Active solar-heating systems use solar panels mounted on the roof. The panels collect and concentrate the sun's energy in a series of tubes containing an antifreeze solution or pumped air. As the antifreeze or air heats up, it is pumped into an insulated storage tank. Fans, controlled by a thermostat, distribute the stored heat through conventional air ducts into the building.

Most areas in the U.S. receive enough sunlight to heat a home at least 60 percent of the year in this manner. In areas that don't receive enough sunlight, backup conventional heating systems are needed.

Domestic hot water can also be created with "active solar water-heating systems." About one million homes in the U.S. get their hot water via this method. Sixty-five percent of all domestic water in Israel is supplied by active solar hot water systems costing the equivalent of $500 per home to install. (See *Solar power, Alternative energy*)

acute toxicity   Refers to the harmful effect a substance has on an organism shortly after exposure to that substance. The harmful effect might be an illness, burns, or death. (See *Chronic toxicity, Toxic waste, LD50, Radiation, Parts per million*)

aerobic organisms   Refers to organisms that require oxygen to survive, as opposed to **anaerobic organisms,** which live without oxygen. (See *Food webs*)

aerobiology   The study of organisms, such as **bacteria** and **algae,** or reproductive cells, such as spores and pollen, that float freely through the air.

aesthetic pollution Aesthetic pollution is difficult to define. Just as beauty is in the eye of the beholder, so too is what someone considers to be environmentally offensive. Aesthetic pollution includes odor, visual, and taste pollution. Some odors may offend almost anyone while others may be unpleasant to only a few. The matter is complicated by the fact that people who constantly smell a certain odor can become oblivious to it after a while. Since aesthetic pollution is defined by the person subjected to it, what is done to resolve it depends on those identifying it. (See *Noise pollution*)

age distribution An important aspect of any **population** is its age distribution. Populations are divided into three groups of individuals: 1) preproductive juveniles; 2) reproductive individuals; and 3) postproductive individuals. Most stable populations, in the wild, have more juveniles than reproductive individuals, and more reproductive individuals than postproductive individuals. This occurs since most species (insects and small animals) have high **mortality** throughout their lives, due to predation (being eaten) or disease. Large numbers of young are produced, but only a small percentage makes it to the reproductive stage and fewer still to the postreproductive stage. For example, less than 25 percent of young cottontail rabbits survive to sexual maturity and many insects species have fewer than 5 percent survive to reproduce. (See *Age distribution in human populations, K-strategists, R-strategists*)

age distribution in human populations Age distribution in human **populations** can be divided into the same three groups as described in **age distribution** for wild populations. However, the number of individuals found in each of the three groups differs.

In a few countries, the three stages are relatively similar in size, so the population tends to remain constant, or even decline. When the majority of the population is in the preproductive stage, as in Mexico, Morocco, and many **less developed countries (LDCs),** a long-term population increase is expected. This is called an "expansive" population profile. If a large percentage of the population is in the reproductive stage, a short-term baby boom is expected, as occurred in the U.S. between the late 1940s and mid 1960s. Once the baby boom has passed through the reproductive stage, the population profile becomes "constrictive," resulting in slower growth. (See *Carrying capacity*)

Agenda 21   One of five documents developed at the **Earth Summit** in Rio de Janeiro in June 1992, it is the largest document, containing over 900 pages in 40 chapters. One of the concerns of Agenda 21 is how the many projects needed to ensure our planet's survival will be financed. Some of the main points discussed in this document are: poverty, changing consumer patterns, population, human health, policy-making for sustainable development, protecting the atmosphere, hazardous wastes, safeguarding the oceans' resources, and promoting environmental awareness. The document was adopted by consensus.

Agent Orange   Agent Orange was a commonly used mixture of two **herbicides.** The drums containing this substance often had bright orange strips, providing its name. During the Vietnam War, over 100 million metric tons of Agent Orange and other herbicides were dumped in Southeast Asia to defoliate areas to expose the enemy. From 1965 to 1970, when the U.S. discontinued use, about 50,000 U.S. military personnel and an unknown number of Vietnamese were exposed to the substance.

Agent Orange was found to be contaminated with the highly toxic substance **dioxin.** Exposure to the contaminated herbicide was suspected as the probable cause of illnesses that developed in many war veterans. In 1984, court action resulted in the manufacturers compensating the victims. (See *Pesticides*)

aggressive mimicry   A form of **mimicry** in which a **predator** mimics a nonpredatory organism. This tactic deceives the **prey** into dropping its guard, making it vulnerable to attack.

air pollution   There are five primary air pollutants: **carbon monoxide, hydrocarbons, nitrogen compounds, particulate matter,** and **sulfur dioxide.** The major source of the first three pollutants is the automobile, which is called a "mobile" source of air pollution. Burning fossil fuels such as coal and oil to generate electricity also contributes to air pollution and are called "stationary sources." In addition to the primary pollutants, **secondary air pollutants** form when the primary pollutants react with each other in the presence of sunlight. Ground-level **ozone** and lead, produced by automobile emissions, also play a major role in air pollution.

When pollutants are released into the air, they are mixed, diluted, and circulated around the globe. Densely populated areas produce large amounts of air pollutants in small regions, making it difficult to dilute and more dangerous to breathe. Local weather conditions create periods of intense air pollution in urban areas, such as **thermal inversions** and **dust domes.** Global wind currents move air across the earth's surface, accumulating air pollutants as it goes. In the U.S., westerly winds carry air pollutants from west to east, further increasing the pollutant concentration.

Until recently, air pollution meant outdoor air. Today, studies have found **indoor pollution** to be a serious problem with a different set of sources and associated problems. (See *Acid rain, Photochemical smog, Ozone depletion, Baubiologie*)

airplane pollution   Airplanes burn **fossil fuels** and cause **air pollution** similar to that of automobiles. The EPA and International Civil Aviation Organization developed airplane emissions standards that are enforced by the Federal Aviation Association (FAA). Just as automobile emission standards have become more stringent, so too have airplane emissions. Airplane fuel accounts for only 8 percent of all the fuel burned for passenger transportation, but more energy is used per person traveling by air than any other mode of transportation.

alar   Alar is a substance sprayed on apples to modify their growth. It postpones the fruit's dropping from the tree, which enhances its color and shape and extends the apple's storage life. In 1989 a public panic arose when it was announced that alar might be **carcinogenic.** This was especially worrisome to parents, since children drink large volumes of apple juice. The risks involved were and still are highly debatable and typify some of the problems involved in environmental **risk assessment.** The facts show that alar causes cancer in laboratory animals, but it has not been established to cause cancer in humans. Few chemicals, however, can be directly linked to cancer in humans. (See *Pesticide dangers*)

alarm pheromone   Refers to a chemical substance released by members of a species to warn other members of the same species of danger. Often used by insects such as ants and aphids. (See *Pheromone, Insects*)

albedo   When energy from the sun enters the earth's atmosphere, one of three things happens. Roughly 35 percent is reflected away by dust particles and clouds. This reflected energy is called albedo. About 15 percent of the energy is absorbed by the atmosphere and the remaining 50 percent reaches the earth and is called insolation.

A theory that was popular a few decades ago proposed that increased pollution in the atmosphere would increase the amount of energy reflected away from earth (albedo), resulting in an overall cooling of our planet and the possible advent of another ice age. However, scientists today are more concerned about the energy that reaches the earth (insolation) and the **greenhouse effect,** which results in increased temperatures. (See *Global warming*)

algae   Algae are primitive aquatic plants ranging from microscopic single-celled organisms to large multicelled plants such as seaweed. Algae is of great importance in many **aquatic ecosystems** since it fills the role of the **producers.**

algal bloom   Refers to a sudden and dramatic increase in the density of **phytoplankton** in a body of water. Algal blooms often occur during **eutrophication.** (See *Standing water habitats, Red tides*)

Allen's Principle   Refers to the concept that warm-blooded animals in cold climates have shorter appendages, such as ears and tails, than in warm climates. Less surface area (on the shorter appendages) results in less heat loss. For example, northern species of rabbits have shorter ears than their southern counterparts. (See *Natural selection, Speciation*)

alley cropping   Alley cropping is a **soil conservation** method in which crops are planted in rows (alleys) between other rows of trees or shrubs. This reduces **soil erosion.** The trees or shrubs can be harvested along with the crop, for fruit or wood. (See *Organic farming*)

Alliance for a Paving Moratorium   The goal of this organization is to halt the environmental, social, and economic damage due to endless road building. Members believe a paving moratorium would limit the spread of population, redirect investment to inner cities, and revitalize the economy. They try to save **wetlands,** farms, and forests from becoming paved in the "name of progress." Write to Alliance for a Paving Moratorium, P.O. Box 4347, Arcata, CA 95521. (See *Urban sprawl, Urbanization, Farmland lost*)

alluvium   Refers to the accumulation of particles such as sand and silt that are carried downriver and deposited along river banks or at the mouth of the river, in areas such as deltas and flood plains. Regions with large amounts of alluvium deposits are considered some of the world's most fertile. (See *Soil, Aquatic ecosystems*)

alpine tundra   See *Tundra.*

alternative energy   Fuels that can replace our dependence on **fossil fuels** (oil, natural gas, and coal) are considered alternative energy sources. Alternatives will become necessary since the combustion of fossil fuels, especially from our automobiles and electric power plants, have side effects that damage our environment, including **air pollution, acid rain,** and **global warming.** In addition, fossil fuels are finite and will someday become exhausted, probably within the next few hundred years.

Alternative energy sources include **nuclear power** and renewable energy. **Renewable energy** sources are considered inexhaustible even if continually utilized by man. They include: **solar power, wind power, hydroelectric power, geothermal energy,** and **biomass energy.**

anaerobic organisms   Refers to organisms that do not require oxygen to survive, as opposed to **aerobic organisms,** which do. Some bacteria are anaerobic and are partially responsible for the decomposition of organic matter (dead plants and animals). (See *Decomposer, Food webs*)

ancient forests   Refers to forests that have never been harvested and therefore contain ancient trees, some 700 or more years old. A typical stand of trees in an ancient forest contains 250-year-old trees with trunks over 20 feet in diameter. Almost all of the ancient forests that once existed on private lands have been logged. Today's standard logging practices cut down trees in a specific area every 60 years, meaning ancient forests, once cut, will never return. Most of the remaining ancient forests are found in the national forests and parks.

The few remaining ancient forests, also called "old growth" forests, are found in the Cascade Range of northern California, western Oregon and Washington, and southeast Alaska. About 2.3 million acres remain, but less than one million acres are designated as wilderness areas and therefore protected from logging. Over the past few years, the **Forest Service** has allowed 60,000 acres of trees over 200 years old to be cut down annually.

These trees are mainly cedar, Douglas fir, western hemlock and

Sitka spruce. Most environmentalists believe all remaining ancient forests should be protected and preserved as natural monuments.

Animal Damage Control program (ADC)   In 1931 Congress created the Animal Damage Control program within the Department of Interior. The Fish and Wildlife Service is responsible for carrying out this program. Its purpose is to destroy "animals injurious to agriculture, horticulture, forestry, animal husbandry, wild game animals, fur-bearing animals, and birds." ADC hunters and trappers kill almost five million animals each year to fulfill this purpose. About half a million of those killed are mammals, including bears, beavers, deer, and mountain lions. Many environmentalists believe the ADC to be a misdirected, misguided attempt to resolve animal/human conflicts simply by killing without any scientific basis or purpose. The ADC's budget is over 20 million dollars per year, which is far higher than any damages perpetrated by the predators being killed. (See *CITES*)

animal manure   Animal manure is used as an ***organic fertilizer.*** It adds nitrogen to the soil, improves ***soil texture,*** and encourages the growth of beneficial ***soil organisms.*** Animal manure use has diminished in the U.S. since animals are no longer raised on the same farms that grow the crops requiring the fertilizer. The need to transport animal manure makes it too expensive for extensive use.

annual plants   Refers to a plant that lives its entire life, from germination through seed production and death, in one year. (See *Succession, secondary*)

Antarctica   Antarctica is the coldest, windiest, and highest continent on the planet, covering 1/16th of the earth's surface. It is covered in ice sheets reaching over 15,000 feet deep and contains about 75 percent of all the world's ***fresh water.*** Temperatures can drop to minus-128°F. One hundred million birds breed each year in Antarctica, and 100 species of fish and mammals—such as porpoises, dolphins and whales—live in its harsh habitat. Antarctica is considered by many to be the last great wilderness on the planet. (See *Biomes, Ice ages*)

anthropocentric   Refers to interpreting the actions of organisms in terms of human values; for example, "The bird must be disappointed that it didn't catch the worm."

anthropogenic stress Refers to the effects human intervention has on other organisms. Although interaction between organisms is universal, the impact of human intervention is usually extraordinary in scope and magnitude. We inflict uniquely "human-caused" stress on the planet. Just a few examples of the anthropogenic stress are *air pollution* and the destruction of habitats as seen in *deforestation* and *desertification.* (See *Biogeochemical cycles, human intervention in; Running water habits, human impact on*)

aposematic coloration Refers to coloration or structures on an organism that warn a *predator* of danger. For example, the distinct stripe on a skunk is aposematic. (See *Mimicry, Directive coloration*)

appliance recycling There are many appliance recycling operations in the U.S. and Canada. They disassemble old refrigerators and recycle various parts and substances. The coolants are siphoned and stored, and metal and glass are separated and shipped to recycling centers. Toxic substances are sent to special incineration centers.

Most are privately owned and contracted by utility companies that want to reduce the number of old, rarely used refrigerators still plugged-in and drawing power. It also fosters goodwill from their customers. Contact your utility company or state environmental protection office to see if an appliance recycling center operates in your area. (See *Recycling, Plastic recycling, Automobile recycling, Motor oil recycling, Tires, recycling*)

applied ecology A division of *ecology* that deals with environmental problems directly affecting our society. This discipline tries to identify existing and potential problems, separate them from imagined or unfounded problems, and attempts to offer solutions. Applied ecology presents facts and theories that are the tools to protect and possibly save our planet from ourselves. (See *Risk assessment, Ecological studies, Ecological study methods*)

aquatic ecosystems About 71 percent of our planet's surface is covered with water providing habitats for many aquatic *ecosystems.* Five factors dictate what kind of ecosystem can exist in water: 1) salinity (the concentration of dissolved salts); 2) depth of sunlight penetration; 3) amount of dissolved oxygen; 4) availability of nutrients; and 5) water temperature. Bodies of water with high concentrations of dissolved salts are called **marine ecosystems,** and those with low levels are **freshwater ecosystems.**

aquifers In the U.S. about half of the drinking and irrigation water comes from underground aquifers. Aquifers are not actual bodies of water, as many people think, but large areas of permeable rock, gravel, or sand that are saturated with water much like a soaked sponge. Aquifers can cover a few square miles up to thousands of square miles, like the Ogallala aquifer, which stretches from South Dakota to Texas.

There are two types of aquifers: unconfined and confined. Unconfined aquifers are located near the surface and are replenished with water directly from the surface in a process called infiltration. The water in these aquifers is not under pressure, so wells tapping them require pumps to draw the water up. Confined aquifers are found deeper in the soil and have an impermeable layer of rock above, so water cannot simply infiltrate down to them. Instead, water enters in areas where the impermeable layer reaches the surface. These "recharge" areas may be miles away from portions of the aquifer. Since water within confined aquifers is enclosed, it is under pressure and moves slowly, usually just a few inches per day. Wells that tap confined aquifers draw water using the natural pressure that exists within the aquifer.

Both the quantity and quality of **groundwater** in aquifers have been affected by humans. The amount of groundwater extracted in the U.S. jumped from 30 billion to over 70 billion gallons per day between 1950 to 1985. In 35 out of the 48 contiguous states, more groundwater is being extracted than can be naturally replenished. This depletion, called "water mining," has resulted in water shortages in many portions of the country.

Pesticides, fertilizers, leaking **septic tanks,** and **toxic waste** from **landfills** are contaminating aquifers across the country. Numerous abandoned **hazardous waste** disposal sites and the practice of injecting **toxic wastes** into deep underground wells are also contaminating aquifers.

Aral Sea   Water diversion redirects water from one area to another, usually for the purpose of irrigating crops. The Aral Sea, in Soviet Central Asia, was the fourth largest freshwater lake in the world until the 1920s. That's when the Soviet government decided to divert the lake and its incoming rivers for irrigation purposes to increase the country's cotton crop. An irrigation canal diverted the water 1,300 kilometers (800 miles) away to farmland.

During the late 1930s, the cotton crop was a success, but the Aral Sea was doomed. Today, the Aral Sea has shrunk in surface area by 40 percent and has lost two-thirds of its total volume. What were shoreline towns have become dry wastelands with lopsided ships littering the land. The fishing trade is gone, along with the crops that once grew alongside the banks. (See *Irrigation*)

arboricide   A chemical that kills trees. (See *Pesticides*)

arboriculture   The cultivation of trees. (See *Reforestation, Temperate deciduous forest, Tropical rain forest, Forest Service*)

Arctic National Wildlife Refuge   The Arctic National Wildlife Refuge is a 19-million–acre refuge located in northeast Alaska containing vast numbers of wildlife, including over 200 animal species. Millions of birds nest and breed in this refuge annually. It is visited by a herd of over 180,000 caribou each year that use it as a calving ground. The Congress has considered opening this preserve for oil exploration and development. (See *National Park and Wilderness Preservation System*)

arid   Pertains to **habitats** that receive less than 25 cm (10 inches) of precipitation annually and the evaporation exceeds the amount of precipitation; for example, a **desert** is an arid **biome.**

artesian well   Refers to a well that does not require a pump to draw water to the surface. Artesian wells draw water from confined **aquifers,** which are under natural pressure.

asbestos   Asbestos has been used since ancient times for its strength, flexibility, and fire resistance. It is also waterproof and sound-resistant. All of these advantages made it a popular building material and insulation during the 1950s through the 1980s.

Asbestos is dangerous when fibers are released and become airborne; this is why old asbestos insulation, which deteriorates, poses such a threat. Cutting, scraping, or sanding materials containing asbestos also releases these fibers and poses a danger. Asbestos fibers, once inhaled, become lodged in the lungs—probably for life. Asbestos is called "the silent killer," since it remains in the lungs for decades before causing disease. These fibers have been conclusively linked to scarring lung tissue and mesothelioma, a rare lung cancer. Due to health dangers, asbestos has been banned and is to be phased out completely by 1997.

Asbestos is, however, still found all around us. It is used in many household appliances such as ovens, toasters, and dishwashers. It is also found in wallboards, and joint and spackling compounds produced before 1970, shingles and sheet floorings in homes built in the 1950s, and is still used in floor tiles sold today.

Asbestos removal is a job for professionals and should not be attempted as a do-it-yourself project. Most states require asbestos removal professionals to be licensed. Your state environmental protection agency could refer you to licensed professionals. (See *Indoor pollution, Respirable suspended particulates*)

aseptic containers   Aseptic containers, also called juice or drink boxes, have become popular alternatives to conventional beverage containers. For the consumer, they have many advantages but many environmentalists feel there are also disadvantages. The advantages include the fact they are sterile, resist breakage, and don't have to be refrigerated. Since they have very little superfluous packaging material, they minimize the amount of waste produced. A full container is composed of only 4 percent packaging materials and 96 percent drink.

The major problem with these products is that they are not readily recyclable. These boxes contain layers of paper (70%), polyethylene plastic (24%), and aluminum (6%). **Recycling** programs for these boxes are few and the ability to establish markets for the recycled materials is unknown since the material cannot be reused to make new aseptic containers. There is also concern that this product may set back advances made in recycling glass and plastic beverage bottles. The market for these recycled materials already exists since they can be used to remanufacture more beverage bottles. (See *Plastic recycling, Source reduction*)

asexual reproduction   Refers to reproduction that involves only one parent and does not involve sex cells, as opposed to **sexual reproduction.** Asexual reproduction can occur by simple division, called fission, which occurs with **bacteria. Fungi** reproduce asexually by producing spores.

Association of Forest Service Employees for Environmental Ethics   The AFSEEE is devoted to changing the current U.S. **Forest Service**'s values to reflect a greater ecological understanding. Members include current, retired, and former employees of the U.S. Forest Service. Their strategies are to provide an open forum for expression of the facts about public land management, provide a support system for Forest Service employees, and educate the public and Forest Service employees on effective ways to practice good land management. They produce a newsletter that voices these beliefs and urge members to speak out about their concerns. Write to AFSEEE, P.O. Box 11615, Eugene, OR 97440.

atmosphere   Refers to the mixture of gases, commonly called the air, that envelopes the earth. Excluding moisture, it is composed of about 79% nitrogen, 20% oxygen, .035% **carbon dioxide,** and a few trace amounts of others such as argon. The lower portion of the atmosphere is called the troposphere, which is where our **air pollution** problems exist. The next higher level is the stratosphere, which is where the **ozone** layer is located. (See Ozone depletion, Greenhouse gases, Carbon cycle)

aufwuchs   Refers to a community of plants and animals living on or around a submerged surface such as a rock or plant stem in a lake or pond. The **dominant** organisms are usually algae, with many insects living in close association. (See Aquatic ecosystems)

autecology   The study of individual organisms or a single species, it concentrates on how an organism's characteristics allow it to survive (or not survive) in certain habitats. Besides studying an organism's anatomy, it also uses sophisticated instrumentation to analyze relations between an organism and its environment at the molecular and chemical level.

An obvious topic of study would be why and how one organism lives in fresh water, while another lives in salt water. More recent studies have resulted in discovering that plants in differing habitats use different types of photosynthesis. C3 plants (which photosynthesize a 3-carbon molecule) are found in all aquatic and most terrestrial habitats. C4 plants (which photosynthesize a 4-carbon molecule) are found only in hot, arid environments. (See Synecology, Ecological studies)

automobile fuel alternatives   Compared to 25 years ago, the U.S. has made great strides in lowering auto emissions, a major contributor to **air pollution,** the **greenhouse effect,** and **ozone depletion.** Unfortunately, even though less is coming out of the tailpipe, there are many more tailpipes on the road today. Therefore, the problem is still getting worse, not better. The U.S. transportation sector uses one million more barrels of oil per day today than it did back in 1973. Federal and state laws demanding improved fuel efficiency are forcing car manufacturers to look for alternatives to conventional gasoline automobiles.

Alternative fuels, including the electric car, offer hope for the future. Compressed natural gas (CNG) and alcohol alternatives, such as methanol and ethanol, reduce emissions. Even though electric vehicles produce zero emissions, they actually displace the source of pollution since electric power plants must still generate the electricity to charge the cars. This, however, will still result in substantial reductions in emissions.

Alternative vehicles using alcohol fuels such as ethanol, methanol, or compressed natural gas will require a distribution network to supply the fuel—something not likely to occur for many years. Electric cars, however, need nothing more than a home electric outlet. Major automotive manufacturers are readying electric cars for production today which should become available soon. Electric cars will have their limitations. Even with technical advances they still have a limited range compared with conventional cars and will probably have steeper price tags, at least at first. The first electric cars will have a range of about 120 miles and then need to be recharged overnight. The car weight will be slightly greater due to the battery, but the acceleration should be similar. (See *Alternative energy, Automobile recycling*)

automobile recycling   Each year about nine million cars are discarded in the U.S. The federal government and auto makers are looking for ways to recycle automobile parts. Executives at the big auto makers believe it is only a matter of time before people will want to buy "green" cars, meaning they will take into consideration how much of a car can be recycled, along with its gas mileage and other features.

The bulk of a car (about 75 percent) is metal and can be extricated from a junked car with shredders, magnets, and other devices. About 43 percent of this metal is now recycled and used by steel makers to make new products, including new cars. A relatively new industry called mini-mills removes the metals from autos (and large appliances) and prepares them for the steel manufacturer.

The rest of the car is called "fluff," which consists of various types of plastics, glass, and other materials. Standardizing the plastics used in autos will enable these materials to be recycled. Nissan, for example, is recycling rubber car bumpers into air ducts, foot-rests, and new car parts. Some German and Japanese car companies have built auto "disassembly plants" to take apart and recycle car components. (See *Recycling, Appliance recycling, Tires, recycling*)

autotrophs See *Producers.*

avifauna Refers to all the birds found in an **ecosystem.** (See *Flyways*)

baby boomers Beginning in 1945 and continuing through the early 1960s, there was a surge in births in the U.S. The birth rate went as high as 3.7 in 1957 from a low of 2.1 in 1937. This boom skewed the **age distribution** of the U.S. population by adding 75 million individuals in a short period of time. Today, almost half of all adults in the U.S. are baby boomers. As these individuals grow older, the age distribution will dramatically shift. The median age of the U.S. population in 1970 was 29, but today it is over 33. It is projected to reach 36 in the year 2000 and 39 by 2010. (See *Carrying capacity, Doubling-time in human populations*)

backshore The area of a beach above the normal high tide level; only covered by water during severe storms. (See *Marine ecosystems*)

bacteria Bacteria are single-celled, microscopic organisms found in many environments in vast numbers. They reproduce asexually by simply dividing (fission). Bacteria play an important role in detritus **food webs** by decomposing organic matter (dead plants and animals) and returning the chemicals to the soil to be reused. Like most organisms, some bacteria require oxygen to survive and are therefore called **aerobic.** Many, however, survive without oxygen and are called **anaerobic.** (See *Nitrogen cycle, Decomposer*)

badland Pertains to an **arid** region such as a **desert,** with little vegetation and noticeable surface **erosion.**

barbecue Charcoal briquettes are composed of coal, limestone, borax, sodium nitrate, and sawdust. Burning charcoal produces the same gases as burning any **fossil fuel.** Burning lighter fluid has been shown to produce numerous compounds harmful both to the environment and to your health and should be avoided if possible. Alternatives include starting your barbecue with newspaper, wax cubes, or hot (electric) irons.

barbless hooks When fishermen release their catch in an effort to save the fish, the likelihood of the fish's survival is dramatically improved when barbless hooks (or hooks that have their barbs flattened) are used. Throwing back a fish that dies due to injuries received from the hook accomplishes nothing. Many trout streams in the U.S. have mandatory **catch-and-release programs** requiring the use of barbless hooks.

barrel of oil A barrel of **oil** contains (is equal to) 42 gallons.

Batesian mimicry A form of **mimicry** in which an edible (nonpoisonous) organism resembles another species that is poisonous. This adaptation protects the edible organism by tricking a **predator** into thinking it is harmful to eat. For example, there is a nonpoisonous snake that resembles the highly poisonous coral snake. Predators leave both alone.

bathyal zone Refers to the level within the open oceans, too deep for photosynthesis to occur, but close enough to the surface for some light to filter through. This region between the euphotic zone (where photosynthesis occurs) and the abyssal zone (where no light penetrates at all) is often called the twilight zone. It ranges from 200 to 1,500 meters (660 to 5,000 feet) in depth. (See *Marine ecosystems*)

batteries Americans use 2.5 billion batteries each year. Disposable batteries are the most common, but rechargeable batteries (also called nickel cadmium batteries or ni-cads) account for about 8 percent of the total. Over the next few years, rechargeables are expected to capture about 20 percent of the market.

There are environmental advantages and disadvantages for both the disposable batteries and rechargeables. Disposables last only a short while and must then be discarded. They contain toxic **heavy metals** such as mercury, which contaminate the soil and **groundwater** as they leach out of **landfills**. Rechargeable batteries, on the other hand, can be used over and over again, minimizing the amount of waste. However, these batteries contain cadmium, another heavy metal. Cadmium is highly toxic, especially to fish, and can cause kidney illness in those who eat the contaminated fish.

Companies that sell both types of batteries should have collection and recycling programs to reduce the potential dangers to the environment and to humans. (See *Biological amplification, Landfills, Toxic waste*)

baubiologie Baubiologie refers to the biology of a building or more specifically, the impact of a building's environment on the health of its occupants. Baubiologie applies this knowledge to help design and construct healthy homes and workplaces. The concept originated in Germany about 20 years ago and has recently become popular in many European countries, but is still in its infancy in the U.S. The concept is advanced in the U.S. by the International Institute for Bau-Biologie & Ecology, which offers certificate programs, architectural consulting, and seminars. The Institute's address is P.O. Box 387, Clearwater, FL 34615. Call (813) 461-4371. (See *Indoor pollution, Healthy homes, Sick-building syndrome, Building-related illness, Asbestos*)

bay A recess in the shore that is larger than a cove but smaller than a gulf. (See *Marine ecosystems, Neritic zone*)

beetles Beetles belong to specific order of ***insects*** called Coleoptera. There are more species of beetles on our planet than all other forms of life (plant and animal) combined. Of the 1.4 million described species, 750,000 of them are beetles. (See *Bugs*)

benthic organisms Refers to organisms that live on or near the bottom of bodies of water. (See *Aquatic ecosystems*)

best available technology (BAT) Refers to the most "state of the art" technology available for a particular industry. Environmental BAT refers to technology that causes the least harm to the environment. For example, most U.S. pulp and paper mills use the best available technologies, whereas many Canadian facilities do not. Using the BAT doesn't necessarily mean a technology is nonpolluting. It only means it is the best available with existing technology. (See *Industrial water pollution, Air pollution*)

Bhopal Bhopal, India, was the scene of the world's worst industrial accident in 1984. Forty tons of a gas (methyl isocyanate) used to manufacture **carbamate** pesticides leaked from a storage facility at a Union Carbide plant. About 3,700 people died and 300,000 were injured. Many people believe the disaster could have been prevented by an investment of about one million dollars. Litigation between India and Union Carbide continues, but it will probably cost Union Carbide hundreds of millions of dollars before it's over. (See *Toxic waste, Pesticide dangers*)

bicycle Just as **alternative energy** sources must be found to reduce our dependence on **fossil fuels,** so too must alternative transportation be found to replace the automobile with its gasoline-driven engine. The battery-powered automobile may offer a solution in the future, and **mass transit** is a viable alternative today in many cities. In many parts of the world, however, the bicycle is the major form of transportation. There are over 800 million bikes in the world, outnumbering cars two to one.

Bicycling is not a form of transportation found only in poor nations, as many people think, where the populace cannot afford cars. In addition to China, with 300 million bikes, Japan, Denmark, and the Netherlands also rely heavily on pedal power.

Bicycles don't pollute, and they relieve traffic congestion and offer healthy exercise. Since they don't burn fossil fuels, they don't cause **air pollution,** as do automobiles. The gasoline consumed driving a car uses 50 times more energy than the energy you burn up riding the same distance on a bike.

Anyone commuting a few miles or less is a candidate to commute via bicycle. In the U.S., with its 100 million bikes, few are used for anything other than sport or pleasure, which means they don't replace the car for most trips. Only 4 million people commute to work on bicycles in the U.S., as opposed to China, where almost everyone commutes daily on bikes, and Japan, where about 15 percent of the work force ride bikes to work.

Why do some nationalities embrace the concept of cycling as a means of transportation while others don't take it seriously? Pro-bicycling countries encourage cycling by providing extensive bike lanes and roads separated from auto traffic with protective barriers. They have nonauto zoned areas within cities and plentiful bike-parking facilities. Although Americans enjoy riding, as indicated by the numbers who do it for fun, we are limited by the lack of available bike lanes and facilities making riding impractical, if not unsafe, for commuting.

The **Rails-to-Trails Conservancy** is an organization that works with railroad companies and municipalities to acquire and convert abandoned tracks into bike lanes. These trails become immensely popular as soon as they open. Those near cities have become daily commutation routes. The popularity of old railroad tracks converted into bike lanes indicates that many more people would ride to work if there were safe and comfortable places to ride. The bicycle has the potential to become an important alternative to the automobile in the U.S.

bicycling, best cities for   According to *Bicycling* magazine, the top five cities in the U.S. for bicycling are: 1) Seattle, WA; 2) Palo Alto, CA; 3) San Diego, CA; 4) Boulder, CO; and 5) Davis, CA.

bio-ore   Bioremediation uses plants and animals to cleanse contaminated environments. One experimental method uses plants that can absorb high concentrations of contaminates such as **heavy metals** from the soil. The soil is cleansed of the contaminate since it all ends up in the plants. The plants, called bio-ores, theoretically can then be processed like a mineral ore to retrieve the metal so it can be recycled for other purposes. (See *Hyperaccumulators*)

bioaccumulation   Many **pesticides** remain toxic for long periods of time. These "hard pesticides," as they are called, remain on plants, where they are eaten and absorbed into the animal's fatty tissue, where they remain. Accumulation of pesticides (or their breakdown products) within an animal's body is called bioaccumulation. These accumulations can harm the animal or be passed on to a predator that eats it, in a process called **biological amplification.** (See *Pesticide dangers*)

biochemical conversion, biofuels Biomass energy can be produced by either **thermochemical conversion** or biochemical conversion. The latter uses bacteria that live without oxygen and feed on the biomass (plants and animal wastes). These bacteria produce methane gas and carbon dioxide as a by-product—a mixture called **biogas,** which is then used as fuel. This process occurs in nature, but can be controlled in **methane digesters.** These devices use plant matter and animal wastes to produce methane, which is collected and used for heating and cooking. China, India, and Korea have tens of thousands of these digesters in use.

Landfills naturally contain these bacteria and therefore generate methane gas. Some of the newer landfills have pipes that collect the gas from the landfill so it can be used as fuel. Experimentally, animal wastes from large feedlots and water treatment plant sludge are also being used to produce biogas for fuel.

Biochemical conversion can also use yeasts to ferment corn, wheat, or other crops to produce alcohol fuels such as ethanol. Brazil uses sugarcane in this manner to produce ethanol, which fuels over half of all the cars in Brazil. (See *Alternative energy*)

biocide   A chemical that is dangerous to all life. (See *Pesticides*)

biodegradable   Refers to the ability of a substance or product to naturally break down into the basic elements or compounds so they can be reused as **nutrients** by plants. This decomposition occurs when bacteria and other microbes feed on the substance. Organic matter such as dead plants and animals and their waste products biodegrade quickly in nature, which is why forests and other habitats aren't littered knee-deep in dead plants and animal carcasses.

Manufactured products such as plastics, however, do not biodegrade readily, if at all. Products that don't break down must be disposed of in some way. (See *Municipal solid waste, Sewage, Landfills, Incineration*)

biodiversity   Refers to the vast diversity of plants and animals on the planet and implies the importance of all. About 1.4 million organisms have been identified, but there could be ten or even a hundred times that many that have not been identified. Organisms are found everywhere. Habitats include fresh, salt, and brackish waters, the soil, and the air. Organisms are found in the arctic, in deserts, and every **habitat** in between. (See *Biodiversity, loss of*)

biodiversity, loss of   When people speak of the loss of **biodiversity,** they are referring to the exceptionally large numbers of species forced to the brink of extinction due to human activities. Species becoming **extinct** is not a new phenomenon and has happened long before humans ever roamed the planet, but the speed with which organisms are being lost is a major concern.

There are many facets to this impact. There is the intrinsic value of every form of life. Many people believe humans don't have the right to force any organism into extinction. There is the impact on an ecosystem. Losing a single species such as a **keystone species**

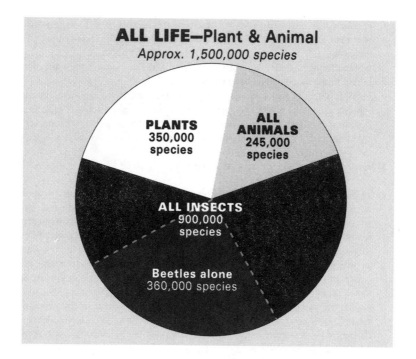

can cause an entire ecosystem to collapse, and even slight changes often have major effects on the entire system.

There are more tangible effects of the loss of biodiversity, including the loss of potentially useful substances such as medicines, crops, and fibers. Possible cures for cancers or other deadly diseases are lost when plants become extinct before they are studied in the laboratory. Some estimate as many as 10 percent of all plants have some medicinal value, many with possible treatments for cancer. It is estimated that when a *habitat* is reduced by 10 percent, 50 percent of the species within that habitat are lost. *Tropical rain forests,* which contain about half of the known species, are being destroyed at a staggering rate.

The most practical way to prevent further loss of our planet's biodiversity is to reduce the loss of wilderness areas by setting aside protected areas. Many environmental organizations such the *Nature Conservancy* are at the forefront of this activity. Also, the federal government must play an important role in protecting the biodiversity of public lands, and balance preserving natural resources with exploiting them. (See *Debt-for-nature swaps, Defor-estation, Ecosystems, Forest Service*)

biofouling Refers to the growth and colonization of plants and animals on submerged surfaces. Biofouling occurs on ship hulls, buoys, wharfs, and almost any other marine surface. Aquaculture nets, which are used by salmon farmers to rear colonies of young fish, are also affected by biofouling. When these nets become colo-nized, nutrients and oxygen cannot get in and waste products can-not get out of the nets, resulting in the death of their inhabitants.

Until 1987, aquaculture farmers used organo-metal paints ap-plied to the net surface to prevent biofouling. These substances are harmful to the surrounding environment and were banned. A new generation of antifoulants are nontoxic and make it physically, as opposed to chemically, difficult for organisms to attach themselves to the net. (See *Water pollution, Hydroponic aquaculture*)

biofuels There are many **alternative energy** sources that can replace **fossil fuels. Biomass energy** is one of these alterna-tives. There are three forms of biomass energy: **biomass direct combustion,** where biomass (organisms and waste) are burned directly for energy; **plant-oil fuel,** where oils naturally produced by certain plants are refined and used as fuel; and biofuels, where biomass is converted into a fuel (biofuel) that is then burned for energy.

Biofuels are produced in two ways: by **biochemical conver-sion,** which uses organisms such as bacteria; and by thermochemi-cal conversion, which uses heat. Biochemical conversion creates fuels such as ethanol and methane (a biogas) while **thermo-chemical conversion** creates fuels such as methanol and syn-thetic natural gas (syngas).

Biofuels are not currently cost competitive, but could become so with advances in technology and increases in the cost of oil. Biofuels could replace one-third of the U.S. consumption of fossil fuels, which would dramatically reduce **greenhouse gases.** There are concerns about cultivating energy crops, which might compete with food crops. Therefore, biofuels produced by waste products, such as **municipal solid waste** or sewage **sludge,** hold the most hope for the future.

biogas Biogas is a mixture of 40% carbon dioxide and 60% meth-ane, which is produced by **anaerobic** bacteria feeding on plants and animal waste. Biogas can be produced under controlled conditions in **methane digesters,** and used as fuel. (See *Biomass energy, Biofuels*)

biogeochemical cycles Refers to the cycling of chemicals essential to life (**nutrients**) between the abiotic (nonliving) and biotic (living) parts of the **biosphere.** Chemicals are taken in from the soil, water, or air by organisms and used as an energy source. Once in the organism, they are transformed into biologically active substances. These chemicals return to the earth when organisms die and decompose. Some of the elements involved in biogeochemical cycles are carbon, nitrogen, sulfur, and phosphorus. (See *Biogeochemical cycles, gas vs. sediment; Water cycle, Carbon cycle, Nitrogen cycle, Phosphorus cycle*)

biogeochemical cycles, gas vs. sediment Gaseous **biogeochemical cycles** occur rapidly—usually taking from a few hours to a few days—as opposed to sedimentary cycles, which take thousands or millions of years to complete. Gaseous cycles move chemicals back and forth between the air or water and organisms. Examples include the oxygen and **nitrogen cycles.**

Sedimentary cycles include the solid earth in the cycle. The sulfur and **phosphorus cycles** are examples. Both gaseous and sedimentary cycles appear to be modified by human intervention. (See *Biogeochemical cycles, human intervention in*)

biogeochemical cycles, human intervention in It is difficult to tell if **biogeochemical cycles** taking millions of years are affected by human intervention. However, changes in short-term cycles are more obvious. The **carbon cycle,** for example, is influenced by extensive burning of **fossil fuels** (oil, coal, and gas) as the world's primary energy source. Burning fossil fuels, like all organic matter, releases carbon dioxide into the atmosphere. The concentration of carbon dioxide appears to be increasing annually. This increase is enough to cause changes in how the atmosphere traps heat and contributes to the **greenhouse effect.**

In addition to burning fossil fuels, vast tracts of forests are being burned (**slash-and-burn cultivation**), adding still more carbon dioxide. Not only does forest burning add carbon dioxide to the atmosphere, but it obviously destroys the trees and other plant life so they can no longer remove carbon dioxide from the atmosphere during **photosynthesis.**

The **nitrogen cycle** has also been modified by human intervention. When fossil fuels burn, nitrogen compounds are released into the air, which reacts with water vapor to create *acid rain.* In some ecosystems, the nitrogen cycle is affected by localized release and subsequent accumulation of fertilizers and livestock wastes in streams. These substances contain high nitrogen concentrations, resulting in the *cultural eutrophication* of ponds and lakes. The *phosphorus cycle* has also been modified by the extensive use of fertilizers and the dumping of wastes into bodies of water.

biogeochemistry Biogeochemistry is primarily concerned with how elements such as carbon, nitrogen, and phosphorus are used by organisms and the impact this has on the chemical composition of the earth. This is closely related to the study of *ecology,* which is more concerned with how the chemical composition of the earth impacts life. (See *Nutrients, Biogeochemical cycles, Photosynthesis, Respiration, Carbon cycle, Nitrogen cycle*)

biological amplification Organisms that feed on plants or take in water that has been contaminated with pesticides can accumulate the pesticide in their tissues in a process called *bioaccumulation.* If these animals are eaten by predators, the substances are passed along. As organisms higher up the food chain feed on contaminated individuals, the concentration of toxic substances increases dramatically in a process called biological amplification.

Biological amplification has been demonstrated in many food chains and numerous toxic substances, but the best-known studies have been done on *DDT* in the Long Island Sound. Water in the Sound was found to contain .000003 ppm (parts per million) of DDT. The plankton living in the water bioaccumulated to a concentration of .04 ppm. Small fish feeding on the plankton had accumulated a concentration of about 0.3 ppm. Larger fish that preyed on the small fish had levels of 2.0 ppm. Predatory birds that fed on these fish, such as the osprey, had levels of 25.0 ppm. This is 10 million times greater than originally found in the water!

Biological amplification can affect any organism. Since humans feed at the high end of *food chain*s, significant concentrations of many toxic substances have been found in our bodies. (See *Pesticide dangers, Breast milk and toxins*)

biological control Before chemicals became the standard method of controlling insect pests, we successfully used natural methods. These natural methods have been advanced with science and technology and are viable ways of controlling pests without chemicals.

Biological control uses populations of **parasites, predators,** and **pathogens** to control pests. Successful implementation of these methods goes back over 100 years when the cottony-cushion scale in California was successfully controlled by releasing the vedalia beetle, a predator. About 300 biological-control success stories have been documented throughout the world.

One of the most common biological control methods is a bacteria called BT *(Bacillus thuringiensis),* which kills many pests. It can be found along with other pesticides in stores. This bacteria comes in different strains, each targeted for different pests. Some strains kill mosquitoes and others are used to control the caterpillars of leaf-eating moths such as the gypsy moth.

Introducing large numbers of ladybird beetles (ladybugs) is another common biological control method. The larval stage of the beetle devours enormous numbers of aphids and scale insects. Many small parasitic wasps (harmless to humans) parasitize immature forms of pest insects, killing them before maturing. Some small wasps, for example, lay eggs in the caterpillar of the tobacco hornworm. The young wasps hatch within their host (the hornworm egg or caterpillar) and feed on the caterpillar's insides, killing it. The well-fed wasps then emerge from the carcass. (See *Integrated pest management, Sex attractants, Natural pesticides, Insect sterilization*)

biological control methods   There are three types of **biological control:** importation, conservation, and augmentation. Importation involves identifying in some remote area, a natural enemy of the pest. The natural enemy is then imported and released in the region containing the pest. The imported insect must be able to survive in the targeted area. Once released, these natural enemies thrive on their own, destroying the pest. In the U.S., the alfalfa weevil has been successfully controlled using this method.

Conservation involves finding natural enemies of the targeted pest locally. Once a natural enemy is found, techniques are implemented to help it prosper and flourish. This includes discontinuing use of any chemicals that are harmful to the natural enemy and using cultural control methods, such as plowing and irrigation techniques that would aid the natural enemy's ability to thrive.

Augmentation involves introducing large numbers of **parasites** or **predators** of the pest. These natural enemies are mass-reared specifically for this purpose. This differs from importation and conservation because the natural enemy isn't expected to reproduce and survive indefinitely as a deterrent. Small parasitic wasps are often used for augmentation control. They lay their eggs into the larval stages of the pest, preventing it from maturing. (See *Integrated pest management, Insect sterilization*)

biological oxygen demand (BOD) Refers to the amount of oxygen needed by microorganisms (usually bacteria) to break down organic material in a specific water habitat over a specific period of time. The more organic matter present, the greater the BOD. BOD is often used as an indicator of water quality. (See *Eutrophication, cultural; Aquatic ecosystems, Water pollution*)

biological pesticides Biological **pesticides** are alternatives to synthetic pesticides, which pollute the environment, contaminate our **groundwater,** and become concentrated in our foods. In addition, many insect pests have become resistant to these synthetic chemicals by **natural selection.** Biological pesticides include the use of many microbes, such as bacteria, viruses, protozoans, and fungi, that attack and destroy certain pests (mainly insects). The sale of biological pesticides has increased almost fivefold over the past five years. (See *Biological control, Pesticides*)

bioluminescence Some organisms, such as the firefly, can convert chemical energy into light. This light is often called "cold light" since very little heat is produced. Only 2 percent of the energy used to generate bioluminescence is lost as heat compared with a typical incandescent light bulb, which loses over 96 percent of its energy as heat, which is why it is so hot to the touch. (See *Insects, Radio-wave light bulb*)

biomass Biomass refers to the dry weight of organisms. It is commonly used in two ways. First, biomass is used to describe the quantity of certain groups of organisms (called trophic levels) that exist in an ecosystem. For example, the biomass of **producers** (green plants) in an ecosystem is far greater than the biomass of primary **consumers.** In other words, by weight, there are more plants than there are animals in a typical ecosystem.

Biomass is also used as an energy source. For example, burning wood or converting plants into a **biofuels** are ways of using **biomass energy.** (See *Energy pyramids, Net primary production*)

biomass direct combustion ***Biomass energy*** is an alternative to burning ***fossil fuels.*** Biomass direct combustion is one of three ways of using biomass energy and refers to burning biomass to create heat, steam, or electricity. Almost any type of ***biomass*** (dry organic matter) can be burned, the most common being wood. About 5 percent of American homes use wood as their primary source of heat. Small wood-burning electrical generation plants also exist, most of them owned by independent power producers associated with the forest industry. Most of these power plants use waste wood produced in lumber and paper mills as the biomass fuel since it is cheaper to burn than to dispose of it.

Other types of biomass are also used for direct combustion. Energy plantations have been established to cultivate fast-growing trees, shrubs, or grasses specifically to provide biomass fuel. ***Municipal solid waste (MSW)*** is often burned in many ***waste-to-energy power plants*** instead of going into landfills. About 8 percent of all MSW in the U.S. goes into these waste-to-energy plants to generate electricity for the surrounding region.

There are environmental advantages and disadvantages to using biomass direct combustion. Substituting biomass fuel for fossil fuels reduces emissions of carbon dioxide, a ***greenhouse gas,*** as long as the plants burned are regenerated for new growth. The new plant growth continues to use up (during photosynthesis) the carbon dioxide emitted by burning the plants; something fossil fuels cannot do.

Air pollution from biomass direct combustion is an important concern. Biomass combustion produces large amounts of ***particulate matter*** that contributes to ***air pollution.*** If municipal solid waste is burned, it might also contain toxic substances, creating a disposal problem. Establishing large energy farms means more environmentally damaging ***monocultures*** and accompanying ***pesticides.*** In addition, there is concern that using land to grow energy plants will compete with food crops.

Although there are some applications for biomass direct combustion, converting biomass into ***biofuels*** shows the most promise as a viable alternative to fossil fuels.

biomass energy Biomass energy is an alternative to fossil fuels (oil, gas, coal) that uses almost any form of organic matter, specifically animal waste and plants, as an energy source. It currently supplies about 4 percent of the U.S. energy needs but has the potential for much more in the future.

Substances currently used for biomass energy include: wood in the form of logs, pellets, or charcoal; plants grown specifically for fuel on energy plantations; or oils from oilseed plants such as rapeseed. Also, almost any kind of combustible waste can be used for biomass energy including: agricultural waste, such as the remaining stalks of crop plants after harvest; timber industry waste, such as wood chips; the combustible part of *municipal solid waste,* such as paper, cardboard, and waste food; and finally animal wastes produced on large livestock farms or in pens.

These biomass substances are either burned directly as fuel (*biomass direct combustion*) or converted to a gas or liquid (*biofuels*) such as ethanol, which can then be burned as fuel. The direct combustion of wood and animal manure supplies about half the energy needs of the *less developed countries.* Municipal solid waste is burned in over 100 waste-to-energy plants in the U.S. Brazil uses surplus sugarcane (the biomass) converted by fermentation into ethanol (a biofuel) to supply about 50 percent of all automobile fuel.

The Biomass Users Network (BUN), founded in 1985, has over forty country members that exchange technologies and ideas to promote the use of biomass energy.

biome   Terrestrial regions of the earth are divided into large ecosystems called biomes, each with distinct combinations of climate, geology, and relatively stable collections of organisms. The two most important factors that determine the types of plants and animals found in each of these biomes are temperature and precipitation. Authorities differ on the number of different kinds of biomes; some descriptions include as few as six or as many as twelve. Eight biomes are listed here: 1) *desert;* 2) *tundra;* 3) *grassland;* 4) *savanna;* 5) *woodland;* 6) *coniferous forest;* 7) *temperate deciduous forest;* and 8) *tropical rain forest.*

Biomes are found at corresponding latitudes and altitudes, since they both can produce the same type of environment. For example, tundra is found in subpolar regions (latitudes near the poles), and also in alpine regions (high altitudes well above the tree line). (See *Zones of life*)

bioremediation   Refers to the process of using organisms to detoxify, absorb, or otherwise render harmless, hazardous or *toxic wastes* found in the environment. Bacteria and plants are being tested to clean up hazardous wastes found in water and the soil. (See *Remediation, Phytoremediation, Bio-ores, Hyperaccumulators*)

biosphere　　The biosphere is that portion of our earth that contains life. This is an incredibly small portion of the planet. Organisms can be found in: a) the lower portion of the **atmosphere** (troposphere); b) on and just below the immediate surface of the land (lithosphere); and c) within bodies of **water** (hydrosphere) and the immediate sediment below. With few exceptions, this places all forms of life no more than a few inches below or a few hundred feet above the earth's land and waters. Exceptions include some microbes and reproductive cells that are occasionally carried by currents high into the atmosphere and some rare forms of **bacteria,** some of which are believed to live in **oil reserves** thousands of feet below the earth's surface.

Biosphere 2　　Biosphere 2 is a privately financed business enterprise that is a cross between a grand-scale science project and an entertainment business. The 3.15 acre greenhouse facility in the Arizona desert, 20 miles north of Tucson, cost $150 million to build. The artificial habitat contains thousands of species of plants and animals and is designed to replicate many natural **biomes,** including a rain forest, a savannah, an ocean marsh, and a desert. In September of 1991, four men and four women were "sealed" inside the facility with the intent of performing ongoing environmental experiments in a closed environment for two years. The wilderness areas naturally recycle gases and purify the waters. Waste would be recycled as fertilizer and a small farm would grow and produce the "Biospherians'" food.

　　At first, the project was attacked as containing more show than scientific substance. The owner of the facility convened a committee of respected scientists to report on the merits of the research aspects of the project. The committee's report stated that the project should spend more time on science and less on business, if it were to provide valuable scientific data and contribute to environmental studies.

biota　　Refers to the living part of an **ecosystem.** The biota is also called the flora and fauna. (See *Abiota, Biosphere*)

Boone and Crockett Club　　The Boone and Crockett Club was founded by Theodore Roosevelt in 1887. Its purpose is to protect **wildlife habitats** and ensure that hunting is practiced in a responsible way. This club helped save Yellowstone National Park from development and pushed legislation that founded the National

Forest System and the National Park Service. By creating the Rules of Fair Chase for responsible hunting, the Boone and Crockett Club serves as one of the nation's advocates of hunters' rights. Write to Boone and Crockett Club, Old Milwaukee, 250 Station Drive, Missoula, MT 59801.

boreal forest   See *Taiga*.

botany   The division of biology involving the study of plants.

bottle bills   First attempts to initiate **recycling** have usually been in the form of a bottle bill. This legislation, passed at the state level, enforces a deposit on bottles so they are returned (for the deposit) and then recycled. Oregon was the first state to implement a bottle bill in 1972. Estimates showed bottle litter down by almost half, two years after the bill became law. So far 10 other states have followed with their own bottle bills. Alternatives to bottle bills are laws that mandate certain types of recycling. (See *Green tax, Disposal fees, Plastic recycling*)

bottom ash   The *incineration* of *municipal solid waste* produces fly ash, which is emitted up the smokestack into the air. What is left are the charred remains that don't go up the stack, called bottom ash. Bottom ash must be disposed of either in *landfills* or by being reincinerated. Since it often contains concentrated amounts of toxic substances, it is often handled as a *hazardous waste.*

boulder   A particle of **sediment** that has a diameter greater than 256 mm (about 10 inches).

brackish water   Refers to water that has some degree of salinity (salt concentration) between fresh and salt water. (See *Aquatic ecosystems*)

breast milk and toxins   One indication of the degree to which toxic substances have permeated our environment is studies showing the existence of these substances in some women's breast milk. The following toxic substances have all been found to occur in mother's milk: *pesticides* such as Chlordane, Lindane, and DDT; *heavy metals* such as lead and mercury; and industrial by-products such as *dioxin.* These substances enter the human body when we eat naturally contaminated foods or are absorbed directly through the skin. (See *Biological amplification, Bioaccumulation, Pesticide dangers*)

breeder reactors   Experimental breeder reactors and conventional nuclear reactors both generate **nuclear power** using **nuclear fission.** Breeder reactors, however, have an advantage over conventional reactors. During nuclear fission in conventional reactors, the fuel rod is used up and must be replaced. This also occurs in breeder reactors, but new radioactive fuel is constantly being created. In this process, a form of uranium that cannot be used as fuel (nonfissionable) becomes useable (fissionable) while the reactor is in operation. This process usually takes about 10 years to complete. This produces a continuous supply of nuclear fuel for nuclear power.

At present there are only a few experimental breeder reactors in existence and research is continuing in many countries. In the U.S., the only breeder reactor, in Tennessee, was never completed and none are scheduled to be built.

Nuclear power generated from breeder reactors have all the same disadvantages as conventional **nuclear reactors.** Breeder reactors, however, are considered more dangerous since the chain reaction is more difficult to control and there is a greater potential for explosions when compared with conventional reactors.

Brower, David R.   David Brower has been called a modern day **John Muir.** He was the director of the **Sierra Club** (founded by Muir) from 1952 to 1969 when the membership grew from 2,000 to 77,000. Brower founded the Friends of Earth in 1969 and the **League of Conservation Voters** in 1970. Later, in 1982, he founded the **Earth Island Institute,** which he now chairs. He was nominated for the Nobel Peace Prize twice during the 1970s and has received numerous honorary degrees. Brower's autobiography is published in two volumes: *For Earth's Sake: The Life and Times of David Brower* and *Work in Progress.*

brown cloud   Large numbers of wood-burning stoves and fireplaces can emit enough **particulate matter** (soot and ash) to produce, under certain climatic conditions, an **air pollution** phenomenon called brown cloud. (See *Thermal inversion, Dust dome*)

Brown, R. Lester   Lester R. Brown began his career as a farmer but started his public career with the U.S. Department of

Agriculture. He is now president of the highly respected **World-watch Institute,** which he founded in 1974, with the help of the Rockefeller Brothers Fund. This research institute is devoted to the analysis of global environmental issues. Ten years after the institute was established, Brown began to publish the annual *State of the World* reports, which are translated into all the world's major languages and have now achieved semiofficial status.

Four years later, in 1988, Brown expanded the institute's publications by launching *Worldwatch*, a bimonthly magazine describing the research of the organization. Mr. Brown is an international speaker and author of many books, such as: *Man, Land and Food, Seeds of Change, By Bread Alone, The Twenty-Ninth Day,* and, most recently, *Saving the Planet.* He is also the recipient of many awards, including the United Nations Environment Prize, Humanist of the Year, and the Robert Rodale Lecture Award.

Btu (British thermal unit) The standard measure for the amount of heat available in a fuel. One Btu is roughly equal to the **energy** released when one wooden match stick is lit.

bug Although often used to describe any **insect,** true bugs belong to a single order called Hemiptera. Included in this group are many insects that feed on plants by sucking the juices out of them, such as milkweed bugs. The ladybug is actually a **beetle.**

building-related illness (BRI) Building-related illnesses involve specific, identified diseases that can be linked to a building's environment, such as Legionnaire's Disease (caused by a bacteria that make the HVAC System their home), or nausea caused by the **outgassing** from building materials. Building-related illness disappears once the cause has been removed from the building. (See *Indoor pollution, Healthy homes, Sick-building syndrome, Heating, ventilation, and air-conditioning (HVAC) systems, Baubiologie*)

bulk elements See *Nutrients, essential.*

bush A small woody plant usually less than 3 meters (about 10 feet) tall. (See *Tree*)

Buzzworm *Buzzworm: The Environmental Journal,* is a well-written and highly informative **eco-magazine** available on newsstands and through subscription. It is published bi-monthly and provides interesting facts and figures as well as full-length coverage of important environmental issues. To order, call (800) 825-0061.

bycatch Some methods of harvesting fish and shrimp result in capturing unwanted species, called bycatch. The bycatch is thrown back into the water, usually dead or injured. Hundreds of millions of pounds of bycatch are killed annually. Common bycatch include dolphins trapped in *purse seine nets,* marine turtles trapped in shrimp trawler nets, and numerous species of fish and mammals trapped in *driftnets* and *gillnets.* In 1990 alone, Japanese drift-netters dumped back into the ocean 39 million fish, 700,000 sharks, 270,000 sea birds and 26,000 mammals, most of which were dead or dying; all considered bycatch. (See *Turtle excluder device*)

C3 plants Plants growing in aquatic and most terrestrial environments build molecules of sugar containing three carbon atoms and are therefore called C3 plants. (See *C4 plants, Photosynthesis, Respiration*)

C4 plants Some plants growing in hot, arid regions produce a 4-carbon sugar molecule, instead of the usual 3-carbon molecule and are therefore called C4 plants. (See *C3 plants, Photosynthesis, Respiration*)

caliology Refers to the study of animal homes such as burrows, nests, and hives. (See *Habitat, Niche*)

Canadian Green Plan The Canadian Parliament passed a massive long-term environmental protection plan and accompanying funding to reduce *municipal solid wastes,* eliminate *hazardous wastes,* and cut back on *greenhouse gas* production and *ozone*-depleting chemicals. It also encourages environmentally sensitive logging, fishing, and farming techniques, and provides for the purchase of lands for parks and protected areas. In addition, the plan budgets monies for *environmental education* and research and outreach programs for *nongovernmental organizations (NGOs).*

canopy Refers to a continuous layer of foliage in a forest formed by the crowns of trees. In *tropical rain forests,* there can be two or three layers of canopy. Each canopy acts as a *habitat* for organisms.

carbamates Carbamates are one of four major categories of synthetic *insecticides* commonly used today to control pests. They are considered "soft pesticides," since they break down into harmless substances quickly after application—usually only a few days or weeks. Carbamates, along with *organophosphates,* kill by disrupting the organism's nervous system. Even though they break down quickly, they are acutely toxic to humans, which means they pose a significant risk to those who apply the chemicals or are in the vicinity of the application. Carbaryl, sold under the brand name Sevin, and aldicarb, sold under the brand name Temik, are examples of carbamates.

carbon cycle   The carbon cycle is one of many *biogeo-chemical cycles.* Carbon is the primary component of all organic matter. The two most important parts of the carbon cycle are: 1) *photosynthesis,* in which carbon (from *carbon dioxide* in the air) and water are converted (using the energy from the sun) into sugar molecules that act as fuel for all living things; and 2) *respiration,* in which these molecules are broken back down to release the energy for the organism to use.

It's estimated that 10 percent of the total amount of carbon dioxide in the air cycles back and forth between the atmosphere and organisms each year through photosynthesis and respiration.

Besides that carbon found in the atmosphere, vast quantities are found in the solid earth (lithosphere). Rocks, soil, and sediment contain carbon. Long-term biogeochemical processes such as weathering and the action of volcanos returns small amounts of this carbon directly into the atmosphere. This sedimentary part of the carbon cycle can take millions of years. (See *Biogeochemical cycles, human intervention in; Greenhouse gases*)

carbon dioxide   Carbon dioxide makes up only .035 percent of the atmosphere (excluding moisture) but plays a vital role in life on this planet. Green plants absorb carbon dioxide during *photosynthesis,* and both plants and animals produce it as an end product of *respiration.*

Carbon dioxide in the atmosphere plays a major role in controlling the earth's surface temperature, since it is the most important of the *greenhouse gases.* The amount of carbon dioxide in the atmosphere has increased over the past years. One study shows an increase from 315 ppm in 1958 to 350 ppm in 1990. Increases in this gas, as well as other greenhouse gases, is believed to cause *global warming.* Increased levels of carbon dioxide are primarily produced from burning *fossil fuels,* such as coal and oil to generate electricity, and gasoline for automobiles. Over 8 billion tons of carbon (as carbon dioxide) are released each year by burning fossil fuels and *biomass* (wood).

Additional increases in carbon dioxide levels are caused by *deforestation.* As forests are cut down, there are fewer trees to absorb carbon dioxide from the atmosphere.

carbon monoxide　Carbon monoxide is one of the five primary components of **air pollution.** It is formed from the incomplete combustion of organic fuels such as oil, gasoline, wood, or solid trash. One of the largest contributors of carbon monoxide into the atmosphere is the automobile. When the car's engine isn't running efficiently, the fuel is not completely burned and carbon monoxide is produced. Inefficient **fossil fuel** power plants also emit large quantities of carbon monoxide into the air.

Carbon monoxide is produced in tobacco smoke and can affect anyone in the area by **passive smoking.** Small amounts of carbon monoxide in minute concentrations can cause headaches, drowsiness, and blurred vision. Cities full of cars and rooms filled with cigarette smoke pose a significant health risk.

carbon tax　See *Green tax.*

carcinogen　Refers to substances that can cause cancer. (See *Carcinogen classification, Toxic waste*)

carcinogen classification　Many people feel that "everything" causes cancer. The statement is not true, but the public's perception is real and to some degree undermines many people's faith in the "system." Much of this perception is created because of the way carcinogens are classified. Some scientists are urging for new classification procedures, not to minimize risks when present, but to provide better scientific evidence and more realistic information about the risks. At present, research is conducted on laboratory animals, the results analyzed, and substances categorized as "known carcinogen," "probable carcinogen," etc.

The major complaints about these procedures include the belief that these categories are too vague and leave little room for explanation. Worst-case scenarios are often used instead of a compilation of all the data. Comments about exceptions or variations that may have been discovered are not taken into account, and contradictory evidence is not included in the final categorization. Reform measures of the carcinogen classification system are being studied by the EPA. (See *Pesticide dangers, Toxic pollution, Acute toxicity*)

carnivore　Animals that eat only other animals are called carnivores, meaning meat-eater. The shark and the dragonfly are both carnivores. (See *Food chain, Consumer, Predator-prey relationship, Energy pyramid*) *carrying capacity* The carrying capacity is the maximum number of organisms that a habitat can support and sustain without degrading each organism's environment. Supporting and sustaining a population means providing enough natural resources such as water, food, and shelter to assure the population's survival. It also involves the ability to eliminate waste products from the organism's environment and the interaction with other organisms in that environment.

The term *overpopulation* is used when the carrying capacity of an area is exceeded, resulting in a degradation of the environment usually followed by a population decline.

Humans have an advantage over other organisms since they can manipulate the carrying capacity of their habitat by changing the way they consume and creating and using technological advances. Attempts to increase the carrying capacity of our planet, by conservation or technology, however, will most likely be futile if the human population continues to increase at its current rate. The

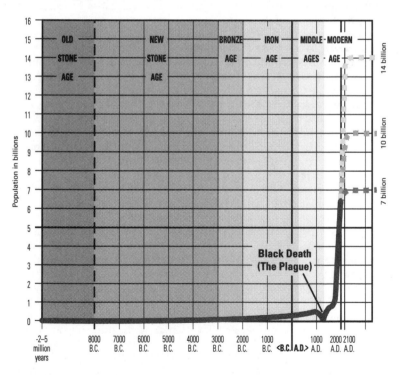

current worldwide "people" population is about 5.4 billion. In roughly 40 years, the population will be about 10 billion if trends hold true. Many scientists believe 8 billion people is the world's carrying capacity, beyond which there would be massive deaths due to starvation and disease, an event already happening on a localized scale in parts of the world. (See *Doubling-time in human populations, Limiting factor*)

Carson, Rachel Rachel Carson, a marine biologist and writer, is best known for her 1962 book *Silent Spring*. In this book she described how pesticides cause long-term hazards to birds, fish, other wildlife, and humans, but provides only short-term gains to controlling the pests. As a result of her work, President John F. Kennedy formed a science advisory committee to investigate her findings. They were soon confirmed, and **DDT** and several other **pesticides** were banned from the U.S. six years later. (See *Rachel Carson Council, Inc., Synthetic insecticides*)

catalytic converter Refers to an emissions control device required on all cars sold in the U.S. It reduces most emissions by over 75 percent compared to cars without the device. (See *Air pollution, Automobile fuel alternatives*)

catch-and-release programs These are programs designed to maintain populations of certain fish in specific areas by requiring or requesting fisherman to throw back fish that meet certain size specifications. For example, anglers in Atlantic Canada are required to release any salmon caught that measures more than 63 centimeters in length since the larger fish are usually females. (Larger females also tend to produce more eggs than smaller females.) This mandatory program is believed to be helping maintain the salmon stock. Catch-and-release programs are enhanced when the anglers use **barbless hooks.**

caulking Caulking and **weatherstripping** a home is considered the easiest, most economical way to save energy and your money. An average house (12 windows and 2 doors) requires 25 dollars' worth of caulking and weatherstripping materials and usually reduces heating and cooling expenses by at least 10 percent. (See *Insulation, Super-insulation*)

Center for Marine Conservation (CMC) The CMC helps to protect marine wildlife and conserve coastal and ocean resources. Some of the projects they are involved with include reducing the use of *driftnets, gillnets,* and *purse seine nets,* and helping to pass a worldwide moratorium on commercial whaling to prevent their extinction. Write to 1725 DeSales Street NW, Suite 500, Washington DC 20036.

Center for Science in the Public Interest (CSPI)
This consumer organization focuses on health and nutritional issues by identifying problems and informing the public to dangers. It regularly reports on deceptive marketing practices, dangerous food additives or contaminants, and flawed science reports, often promoted by some industries. CSPI has successfully obtained restrictions on several suspicious food additives and routinely evaluates the safety of new additives. Due in part to the organization's efforts, major fast-food chains have switched to less saturated fats, the FDA has banned many uses of sulfites (a preservative), and the beer industry has eliminated cancer-causing nitrosamines from its products. CSPI publishes several reports and newsletters. Its address is 1875 Connecticut Avenue NW, Suite 300, Washington DC 20009. (See *Pesticide dangers, Fruit waxing, Factory farms*)

CFC (chlorofluorocarbon)
CFCs contribute to *global warming* and are the primary culprit responsible for *ozone depletion.* CFCs are gases used as refrigerants in air conditioners and refrigerators. They are also used as propellants in aerosol containers. Freon was the original CFC, developed in the 1930s.

In recent years the concentration of CFCs in the atmosphere has substantially increased as they escape from the products mentioned above. In 1976, the concentration of chlorine (from CFCs) in the atmosphere was 1.25 parts per billion. In 1989 it was almost twice that number. Although this might sound like a ridiculously small amount, CFCs are about 15,000 times more efficient than carbon dioxide at producing the *greenhouse effect,* so a little means a lot. About 20 percent of global warming is attributed to CFCs.

An even more important consequence of CFCs, however, is their impact on the *ozone* layer. CFCs destroy ozone and, to make matters worse, prevent it from reforming, which occurs naturally. Once in the atmosphere, CFCs linger for 50 to 100 or more years; so it is difficult to reverse damage already done.

CFC use in aerosol cans has been banned in the U.S. for many years. New substances have been developed to replace them. In 1987, many of the *more developed countries,* including the U.S. and Canada, agreed (in a document called the Montreal Protocol) to freeze production of all CFCs and other pollutants at their existing levels and then begin to reduce the amount by 50 percent by the year 2000. In 1989, over 80 countries signed the Helsinki Declaration, which stated all CFCs and other pollutants would be phased out of use by the year 2000.

HCFCs (hydrochlorofluorocarbons) are now used as a replacement for CFCs. This molecule breaks down more quickly so it has less time to destroy the ozone layer. HCFCs last from about 2 to 10 years compared with CFCs, which last for 50 to 100 years. Alternative solutions are needed, however, that don't cause *ozone depletion* at all.

char   The charcoal-like organic remains of the combustion of biomass. (See *Biomass direct combustion, Incineration*)

Charles A. Lindbergh Fund, Inc.   This organization was founded in 1977, to honor Charles A. Lindbergh and carry on his beliefs: that the future of mankind depends on a balance between technological advancement and environmental preservation. Each year, this fund presents grants for research and educational projects in areas such as aviation, agriculture, arts and humanities, biomedical research and adaptive technology, water resource management, and wildlife management. Write to 708 South 3rd Street, Suite 110, Minneapolis, MN 55415.

chemozoophobous   Refers to plants that protect themselves from being eaten by producing noxious substances.

Chernobyl   The worst nuclear power plant accident occurred in a small city north of Kiev in what used to be the Soviet Union on April 26, 1986, at 1:23 A.M. During a test, engineers violated regulations and turned off most of the automatic safety systems. The test resulted in two major explosions that blew the 1,000-ton roof off the *nuclear reactor* and set fire to the reactor core. Radioactive debris flew into the air and was carried by the wind over much of Europe. Areas over 1,000 miles away became contaminated. Over 135,000 people were evacuated, and the region was secured.

The Soviet Union acknowledged 36 deaths three years later, but many from within the Soviet Union say it was more like 300 dead. Medical experts estimate that between 5,000 and 150,000 people in the region will die a premature death because of the accident at Chernobyl. Recent reports show the number of children with thyroid cancer has soared from one or two cases a year prior to the accident to over 130 in 1991 in regions near the site. In 1992,

Ukraine's chief epidemiologist reported a 900 percent increase in leukemia in those villages closest to the explosion.

The cleanup of the facility has cost over $14 billion and is expected to be much higher before completed. The reactor is now entombed in concrete, but is showing signs of cracking, and some believe it to be releasing radioactivity. (See *Nuclear reactor safety and problems*)

chernozem   See *Soil types.*

Chief Seattle   Some of the most quoted inspirational words about the environment are attributed (often incorrectly) to a nineteenth-century Native American Indian, Chief Seattle. The Chief gave a speech in 1854 (now called the Fifth Gospel speech), which included passages about the environment. Many of the words often quoted by the media, however, did not come from the Chief's mouth. For example, a quote about seeing the slaughter of thousands of buffaloes was actually written by a screenwriter in 1972 who put the words into the Chief's mouth for a movie. (See *Ethics, environmental; Philosophers, environmental*)

chledophyte   A plant that can grow on rubbish heaps, such as on **landfills.**

chlorinated hydrocarbons   Refers to one of the major categories of synthetic **insecticides. DDT** was the first chlorinated hydrocarbon insecticide and probably the best known. Chlorinated hydrocarbons effect an organism's nervous system, usually resulting in death. They can be applied once and last for long periods of time, which is both a blessing and a curse. Since they are so persistent and remain in the environment for 2 to 15 years they need not be continually applied to control a pest.

These chemicals, however, are harmful to many nontarget (nonpest) organisms. Since they last so long, and can be absorbed and stored in an animal's body, they are passed along food chains and accumulate in animals, including humans, in a process called **bioaccumulation** and **biological amplification.** Since they are so persistent and effect many forms of life, most chlorinated hydrocarbons have been banned in the U.S. They are, however, commonly used elsewhere in the world. Other common insecticides in this category are Aldrin, Chlordane, Dieldrin, Endrin, Kepone, Lindane, and Mirex. (See *Pesticides, Pesticide dangers*)

chlorination   Chlorine has been added as a disinfectant to *drinking water* since the early 1900s. Seven out of 10 Americans drink chlorinated water, but some scientists are concerned about its safety. Chlorine reacts with organic matter in the water, such as decaying leaves or grass, and produces substances called trihalomethanes (*THM*), which is believed to be carcinogenic. The EPA regulates this substance by issuing "annual" average amounts that can be present in a municipality's drinking water. This means a municipality can add high concentrations of chlorine on any given day, posing a possible health threat. (See *Sewage treatment, Water treatment*)

chlorophyll   Refers to the green pigment found in plant cells essential for *photosynthesis.* (See *Producers*)

chronic toxicity   Refers to a harmful effect, such as illness or death, after the long-term exposure to low dosages of a substance. (See *Acute toxicity, Toxic waste, LD50*)

circle of poison   Refers to the sale of *pesticides* that are banned in the U.S. but exported to other countries. There, they are used on crops, which are then imported back into the U.S. and sold in American supermarkets. Much of this produce contains pesticide residues that would be illegal if grown and sold in the U.S. (See *Pesticide dangers*)

CITES (Convention on the International Trade of Endangered Species)   CITES was founded in 1973 and is located in Lausanne, Switzerland. Its mission is to monitor and regulate the international wildlife trade. Over 100 countries belong to the organization. It meets every other year to determine which organisms require protection and then implement restrictions. The organization uses three categories: The first prohibits all international trade of an *endangered species.* The second permits some trade, and the third category leaves the regulations up to the countries involved.

Although CITES pass laws that member countries are required to abide by, there are many loopholes and enforcement is difficult. The two major concerns CITES addresses are wildlife habitat destruction and unrestricted wildlife trade, which is a multibillion-dollar-a-year business. (See *Endangered Species Act, Ivory trade, Poaching, Extinct*)

Clean Air Act This is the cornerstone piece of legislation, passed in 1970, designed to improve the quality of America's air. Its goal is to assure Americans that the air they breathe will pose no health risk. The law is enforced by the EPA. In 1990, the act was renewed and amended to include the following new mandates: Most major cities must comply with reduced emissions of primary air pollutants that cause *air pollution* and *smog* by 1999. (Nine of the largest cities were given a few extra years to comply.) It requires reductions in auto tail-pipe emissions and special nozzles on gasoline pumps to reduce *volatile organic compound (VOCs)* fumes.

The law also requires industries emitting any of 189 identified toxic chemicals to use the *best available technology* to reduce these emissions. Limits were also placed on the amount of *sulfur dioxide,* partly responsible for *acid rain,* released from fossil fuel—burning power plants. (See *EPA past and present, President's Council on Competitiveness*)

Clean Water Act This is the cornerstone piece of legislation designed to "restore and maintain the chemical, physical, and biological integrity of the nation's waters." The law, passed in 1972, has resulted in a marked improvement in the quality of our waters. The law mandates each state to adopt water quality standards for all streams and restricts industry and municipal waste discharges and protects wetlands. It is enforced by the EPA. The Water Quality Act of 1987 reaffirmed and strengthened this law. This act was reauthorized in 1992.

A portion of the act responsible for *wetlands* protection has become a political battleground in recent years. Some members of Congress favor a new bill that redefines wetlands and would result in the destruction and development of 80 percent of those wetlands that still exist. Other members back a pro-environment bill that would leave the definition of a wetland up to scientists, instead of politicians. (See *Water pollution, Aquatic ecosystems, No net loss, Estuaries and coastal wetlands destruction*)

clear-cutting Clear-cutting is a forest logging method in which every tree in a region is cut. This is one of the oldest methods of harvesting forests and used globally. Clear-cutting is economical for the logging industry, but devastating to the forest. When large tracts of forest are cleared, the habitat of most animals is removed and the ecosystem for the entire region affected. Clear-cutting, especially on sloped terrain and where regrowth is slow, causes *soil* *erosion,* making it impossible for the ecosystem to recover. Global clear-cutting is believed to be affecting the overall balance of *carbon dioxide* throughout the *biosphere.* The use of clear-cutting in the U.S. national forests has angered environmentalists, who are trying to curtail further destruction.

Clear-cutting relatively small tracts of forests with little slope and rapid regrowth has been used successfully for decades. This "patchwork clear-cutting" removes only patches of forest, but leaves the surrounding growth untouched. This produces less harm to animal habitats, and is less likely to result in soil erosion. (See *Forest Service, past and present; Deforestation*)

climax community   Refers to a final, stable, and self-perpetuating *community* of organisms in a region. (See *Succession, Biomes, Aquatic ecosystem*)

coal   Coal is a *fossil fuel* that supplies 28 percent of the world's fuel. About 65 percent of all coal used is burned in boilers to produce steam which in turn produces electricity. In the U.S. over half our electricity is produced from coal in this manner. Most coal is extracted from the earth by either *strip mining* or *subsurface mining.*

Over 50 percent of the world's coal reserves (identified deposits) are located in China, with 10 percent in the U.S. Coal reserves are expected to last about 220 years if existing demand continues. The projected amount of coal resources (unidentified deposits), however, might last about 900 years with the current demand.

Coal contains high heat content at economical costs, but many environmental problems are associated with its use. Coal is the dirtiest fossil fuel to burn, so sophisticated air pollution control devices are necessary. Burning coal releases pollutants that contribute to *acid rain* and more carbon dioxide than any of the other fossil fuels, increasing the *greenhouse effect.*

Strip mining for coal devastates an area and often results in serious erosion. Even when attempts are made to reclaim the land with grading, return of topsoil, and replanting, the ecosystem never fully recovers its *biodiversity.*

New technologies can burn coal more efficiently and more cleanly. This includes the fluidized-bed combustion method, which will begin replacing old-style coal burners later in this decade. It converts solid coal into gas or liquid fuels called *synfuels.* (See *Nonrenewable energy*)

coal formation   *Coal* began to form 300 million years ago (during the Carboniferous Period) when large regions of the earth were covered with tropical swamps containing dense vegetation. As the fast-growing vegetation died and accumulated under the water, it formed a material called peat, which is the first step in the formation of coal. Peat is composed of about 90% water, 5% carbon, and 5% other substances. The peat was gradually covered by ***sediment.***

Over time, pressure squeezed out much of the water and compressed the peat into lignite coal (also called brown coal), which contains about 40% moisture. With heat from the earth and continued pressure, lignite was transformed into a soft type of coal called bituminous coal, which has only about 3% moisture. With continued heat and pressure, hard coal called anthracite was finally formed.

The entire process took hundreds of millions of years. The products produced during each stage are found today. In some ***less developed countries,*** peat is dried and burned but has a low heat content and makes a poor energy source. Lignite also has low heat content and is not a good source of energy. Bituminous coal has a high heat content and is the most common type of coal. Unfortunately, it has a high sulfur content, making it more harmful to the environment. ***Air pollution*** control devices are used to reduce these emissions. Anthracite has a high heat content and a low sulfur content, making it the most desirable energy source. This coal, however, is in limited supply and expensive. (See *Oil formation*)

coal gasification and liquefaction   See *Synfuels.*

coal gasifier   Almost any fuel will burn more cleanly and efficiently if it is first converted to a gas. When this principle is applied to organic matter (***biomass*** or ***fossil fuels***) it is called ***gasification.*** A promising form of gasification involves using coal gasifiers that burn coal (a fossil fuel) far more cleanly than conventional coal combustion, producing less ***air pollution*** and fewer ***greenhouse gases.*** It also burns more efficiently, getting more energy out of the same amount of coal when compared with conventional coal burning. Even the most efficient coal-burning technologies, however, cause more environmental harm than either ***oil*** or ***natural gas.***

Coast Redwood   The Coast Redwood is the tallest species of tree in the world and one of the oldest. An average mature tree stands 220 feet tall, but some have measured in at 360 feet with a trunk diameter of 10 to 15 feet. These trees can live for over 2,000 years! They are found only in a narrow band along the Pacific Ocean from the southern tip of Oregon to central California. The *Save-the-Redwoods League* has been helping to preserve these stands since 1918.

Coastal Society, The (TCS)   The Coastal Society, chartered in 1975, serves as a forum for coastal resource professionals and others interested in promoting a better understanding and sustainable use of coastal resources. Its goals are to: encourage cooperation and communication, promote conservation and wise use of coastal resources, help government and industry balance development and protection along the world's coastlines, and advance public education and appreciation of coastal resources. Write to The Coastal Society, P.O. Box 2081, Gloucester, MA 01930-2081.

coastal wetlands   Land that is flooded for all or part of the year is called a *wetland.* If it contains saltwater, it is called a coastal wetland (as opposed to *inland wetland*). Coastal wetlands in temperate regions include bays, lagoons, and salt marshes, and all have grasses as the *dominant* vegetation. In tropical regions, coastal wetlands are primarily swamps inhabited by mangrove trees. (See *Estuary and coastal wetland destruction, Wetland*)

coastal zone   See *Neritic zone, Ocean.*

cobblestone   A rounded rock particle between 64 and 256 mm (2.5 to 10 inches) in diameter; bigger than a pebble but smaller than a *boulder.* (See *Soil particles*)

cogeneration   About one-third of the energy produced by a conventional fossil-fuel power plant is converted into electricity; the rest is lost as heat. Cogeneration refers to capturing and using the lost heat, which can be used directly to heat the facility or can be recaptured to generate more electricity. (See *Coal. Alternative energy*)

cold resistant   Refers to the ability of an organism to survive freezing temperatures. Some *insects* produce an antifreeze to allow them to survive the winter.

commensalism   Commensalism is a **symbiotic relationship** in which one organism benefits from the relationship and the other is unaffected by it. Many mosses, lichens, and vines grow on trees in this type of relationship. The classic example is that of a fish called the remora, which attaches itself to sharks with a sucker appendage and feed on the scraps of food left by the shark during feeding. The shark appears to be indifferent to the relationship. (See *Parasitism, Mutualism, Parasitoidism*)

Commoner, Barry   Barry Commoner, a scientist and writer, was one of the first to make the public aware of the tremendous environmental costs of our technological development. Commoner reveals, in his latest book, *Making Peace with the Planet,* how science, politics, the private sector, and public policy all must be examined as a whole if we are to preserve our resources and not waste any more money or time. He currently directs the Center for the Biology of Natural Systems at Queens College in New York City.

community   Refers to a collection of all the **populations** inhabiting an area at a given time. (See *Ecosystems*)

community ecology   See *Synecology.*

Community Right to Know Act   In 1986, Congress enacted this law requiring companies to report their emissions of over 300 chemicals. The law contains many exceptions and exemptions resulting in only one out of about 20 pounds of toxic emissions actually being reported. (See *Toxic waste, Air pollution*)

competition   When two organisms of the same species compete for the same resource, it is called intraspecific competition. The resource might be food such as a plant or prey, or it might be an abiotic (nonliving) factor such as sunlight or water. If the two organisms competing for a resource are of different species, it is called interspecific competition. (See *Predator-prey relationships, Niche*)

compost   Compost is one of three types of **organic fertilizer.** It is a rich soil containing a large amount of decomposed organic matter. Compost is created from wastes such as lawn clippings, leaf litter, food waste, and animal droppings that are mixed with topsoil and decomposed by populations of microbes. These microbes decompose the organic matter, making it available for future plant growth. About 34 percent of the U.S. *municipal solid waste* consists of organic matter that is compostable, but only 1 percent is currently composted. There are facilities that use *sewage* sludge from *water treatment* plants as the organic material to create compost. (See *Composting*)

composting   Composting is the process of converting organic waste into *compost.* Leftover food, grass clippings and leaves, animal wastes, and sewage sludge are all candidates for composting. This organic material is broken down (decomposed) into the essential nutrients for future plants to use. Composting in the U.S. is usually done on a small scale as described here, but some countries and U.S. cities have performed large scale composting using either *municipal solid wastes* or *sewage* sludge.

A typical small or garden type of composting setup consists of a wire frame (that allows air through) about five feet square. A layer of twigs on the bottom lets air circulate. Add a layer of dry organic matter such as leaves or grass clippings over the twigs. Then add your organic wastes (kitchen wastes, animal wastes, etc.). Over each layer of organic waste, place a thin layer of soil. Continue adding layers of waste and soil until the frame is full, or at least three feet high. The pile should be turned over once every two months. Leave a depression on the top to collect water so it seeps in. In 6 to 12 months the organic matter will have decomposed into a rich compost that can be used as fertilizer. The compost is ready when it has turned brown to black and has no unpleasant odor. The ingredients should no longer be identifiable. (See *Composting, large-scale; Biomass energy*)

composting, large-scale   Some European countries, and a few U.S. municipalities have compost plants that recycle organic wastes from *municipal solid wastes* or *sewage* sludge and sell the compost as fertilizer. Even though the compost is sold to generate revenues, most of the cost savings come from reduced landfill fees since there is less waste to dump. These large-scale compost plants agitate and aerate the organic mix to speed the natural *composting* process along. The bacteria that decompose the organic matter thrive with the added air and moisture. What takes many months to decompose in nature takes only a few weeks in these compost plants.

Comprehensive Environmental Response, Compensation, and Liability Act (CERCLA)
See *Superfund.*

comprehensive water management planning
The process of developing, but still protecting, a region's water resources is called comprehensive water management planning. This process often requires the integration of many agencies, organizations, companies, and other groups that use the resource and/or have jurisdiction over it. The water management plan involves data collection and analysis, problem determination, and recommendations for resolutions. It often includes studies that require the continued monitoring of the resource over long periods. (See *Water use, Water pollution*)

computer modeling, ecological Refers to the use of high-speed supercomputers that simulate changes in biological systems so predictions can be made about their impact. The biological systems studied may be a single population, a community, or an entire ecosystem. These models assist in all forms of **ecological studies,** predict what might happen to an ecosystem in the future, and help to develop new ecological theories. (See *Ecology study methods*)

computer networks, environmental Computer networks enable computer users throughout the world to share information. Some of these networks are expressly designed to share and disseminate information about environmental issues. (See *EcoNet, MNS Online*)

computer-produced pollution Although not usually thought of as an industry with pollution problems, computers, by their sheer numbers, have become a focal point of environmental concern. Many computer companies have recently become involved in environmental concerns. The silicon boards that make up the guts of a computer and associated devices such as printers, must be cleaned after production by solvents normally containing **CFCs.** Apple Corporation was the first company to eliminate the use of CFCs for this process. Compaq, DEC, Fujitsu, IBM, and Intel have all announced plans to follow suit.

The corporate standard in personal computer printing is the laser printer, which requires a disposable toner cartridge. About 20 million of these cartridges are shipped and end up in landfills each year.

Apple started a toner cartridge recycling program back in 1991, and other companies have begun to follow suit. Kyocera has produced a new technology printer that doesn't require the conventional toner cartridge at all. The few replaceable parts that do exist are biodegradable.

Old computers need not be scraped since most can be refurbished, upgraded, and at least kept in working order. What is outdated in some places may be considered "high tech" elsewhere. The East-West Education Development Foundation in Cambridge, Massachusetts, accepts donations of old systems, which they ship to Russia and eastern Europe; call (617) 542-1234. Many local charities and schools also accept old computers. (See *Laser printer ozone, Copier toner cartridge, recycled*)

computer software, environmental There are many computer programs available that are designed to entertain, educate, and inform people about environmental topics and issues. The Save The Planet Software Company of Pitkin, Colorado, (303) 641-5035, has computer programs (for DOS and Mac) that provide detailed information about global environmental issues such as ozone depletion and global warming. These programs use graphics, charts, maps, and text to present the information. Chariot Software Group, (619) 298-0202, of San Diego, California, has many environmental programs and games for the Macintosh. (See *Computer networks*)

CONCERN CONCERN is a nonprofit organization that provides environmental information to community groups, schools, individuals, and public officials who are interested in environmental issues. Its primary activity is developing and publishing a variety of community action guides that offer a comprehensive explanation of these issues, such as *global warming,* household waste, and *pesticides*. Write to CONCERN, 1794 Columbia Road NW, Washington DC 20009.

conchology The study of sea shells.

coniferous forest See *Taiga.*

conservation districts Throughout the U.S., there are 3,000 local conservation districts, which help decide the ultimate environmental quality of a town or city. These conservation districts are authorized under state laws to assess environmental problems, set priorities, coordinate and carry out efforts at the local level to address these problems. Each district is governed by either appointed or elected officials. (See *Ecocities, Zoning, land use; Soil conservation, Urban sprawl, Urbanization*)

conservation tillage   "Conventional" tillage farming invites **soil erosion** since it involves turning over the soil in the fall, leaving it bare through the winter, and planting crops in the spring. Conservation tillage methods use special tillers that break up the subsurface, but don't disturb the surface layer, leaving a protective layer of organic matter. Seeds, **fertilizers,** and pesticides are injected through the surface to the layer below. Leaving the top layer intact greatly reduces the amount of erosion.

Conservation tillage is used on only about one-third of U.S. crops. It is estimated that if 80 percent of U.S. croplands used this method, about half of all the soil lost to erosion would be saved. (See *Conservation, Green manure, Organic farming*)

constancy   See *Population stability.*

constant species   Refers to a species that is likely be found in a specific **community.** A constant species is found in at least 50 percent of the samples taken from that community. (See *Keystone species*)

consumer   Organisms can be categorized according to how they obtain energy to survive. Animals that must consume plants or other animals to obtain their energy are called consumers (as opposed to **producers,** which capture energy from the sun during **photosynthesis** to produce their own food). Animals that eat the plants are primary consumers and include cattle, rabbits, and grasshoppers. Animals that eat the primary consumers are secondary consumers and include many predatory birds, fish, and insects. Tertiary consumers eat secondary consumers and include lions, sharks, and hawks. Which types of organisms fill the role of producers and consumers depends upon the type of **biome** or **aquatic ecosystem** in which they are found. These relationships are illustrated in **food chains.** (See also *Energy pyramid, Producer, Decomposer, Scavenger, Predator-prey relationship, Parasite*)

contagious disease   Refers to any disease transmitted by physical contact. (See *Infectious disease*)

contour farming Crops grown on gradually sloping land are prone to **soil erosion.** Contour farming is a **soil conservation** method in which crops are planted at right angles to the slope, instead of up and down the slope. Each row creates a sort of minidam that holds the water (and soil) in place.

Convention on Climate Change, The This document is one of the five documents discussed during the **Earth Summit,** held in June 1992. The issues discussed in this convention were the following: reducing emissions of carbon dioxide, the primary cause of **global warming;** controlling emissions of other greenhouse gases; and the need for financial and technical aid to developing countries now dependent on **fossil fuels.** The main objective was "to stabilize the greenhouse gases in the atmosphere to prevent dangerous interference with the climate system, to ensure that food production is not threatened and to enable economic development to proceed in a sustainable manner."

cool energy The combustion of **fossil fuels** emits **greenhouse gases** that raise global temperatures. **Alternative energy** sources that can replace our dependence on fossil fuels are called cool energy. Cool energy can come from wind, hydro, solar, or biomass (plant) sources. These are all **renewable energy** sources, since they are considered inexhaustible. The amount of research funding for cool-energy sources dropped from about one billion dollars in 1981 to less than one tenth that in the early 1990s.

copier toner cartridges, recycled Most photocopy machines have a disposable toner cartridge. Xerox Corporation has introduced a new copier that uses recycled cartridges. After about 25,000 copies the cartridge is sent back to the company, where it is recycled and returned to the customer. (See *Recycling, Computer-produced pollution, Paper recycling*)

coprophagous Refers to organisms that feed on animal dung. There are many species of dung beetles that live most of their lives in a pile of dung. Some prefer a certain type of dung; cow or horse, for example. Some, such as the tumblebug (which is really a **beetle**), chew off little pieces of dung and roll them into balls. They then roll the balls to a safe location, lay their eggs in the balls, and bury them. When the eggs hatch, the young are protected in the soil and fully stocked with food. (See *Insects, Decomposers, Food web*)

coral bleaching   Refers to the death of coral and the entire ***coral reef*** ecosystem in a region. This may occur for many reasons: ***sewage*** or other waste flowing downstream and into the oceans; silt accumulation from ***soil erosion*** caused by improper logging and agriculture; or from contaminants such as ***pesticides.*** (See *Neritic zone, Ocean pollution*)

coral reefs   Marine ecosystems along the shore (***neritic zone***) contain many unique habitats, such as coral reefs. Coral reefs are found in warm tropical and subtropical regions. The reef is built primarily by tiny coelenterates, which are cylindrical in shape with a pouchlike mouth surrounded by tentacles that capture prey and bring it into the mouth. (Sea anemones and jellyfish are other types of coelenterates.)

These organisms attach themselves to the floor of coastal waters. They secrete calcium, which acts like a substrate for other individuals. As these calcium deposits increase, coral reefs are formed. Only the surface of the reef contains living coral organisms. The rest is just the remaining calcium deposits.

Some coral have a ***symbiotic relationship*** with an algae that lives within the coral's cells. The coral captures its food at night, but during the day, they are nourished directly from the photosynthesizing algae living within their body.

Coral reefs, along with other coastal zone habitats, are among the most productive on earth. They are as productive (***net primary production***) as ***tropical rain forests.*** Thousands of species of fish and other organisms, such as sea urchins, feed on the algae, bacteria, and other microorganisms that abound in coral reef ***ecosystems.***

Cousteau, Jacques-Yves   Jacques Cousteau has explored every major body of water on our planet. In order to explore the world underwater for long periods of time, Cousteau and Emile Gagnan invented the Aqua-lung. He is also credited with developing the first underwater cameras with television capabilities. Through this medium, he has not only been able to study the underwater world but teach everyone about it as well. (See *Coral reefs, Environmental education*)

crop rotation   Crop rotation is a farming practice that prevents the soil nutrients from becoming depleted. Some crops rapidly deplete the soil of nutrients while others replenish nutrients. Crop rotation involves alternating these kinds of crops so the nutrient level in the soil remains stable. Corn, tobacco, and cotton deplete nutrients (especially nitrogen), but legumes such as oats, barley, and rye add nitrogen, by the process of *nitrogen fixation.* Crop rotation also reduces the chance of pest infestations and plant diseases, since the pest populations cannot build up over many years. (See *Organic farming, Soil conservation*)

cruise-ship pollution The **Center for Marine Conservation (CMC)** has reported that over 20 cruise lines have not been following the 1988 worldwide ban on plastics dumping at sea. Plastic litter from cruise ships have been washing ashore in 32 U.S. states and 12 countries, with the majority of the waste coming from four major cruise lines. (See *Plastic pollution, Plastic recycling*)

cryptozoology Refers to the study of small terrestrial animals that live in crevices, under stones, and in *leaf litter,* which includes many small *insects* and crustaceans. (See *Habitat, Niche*)

cullet Refers to crushed glass used for *recycling.*

culling The selective removal or killing of individuals to reduce the size of the population. (See *Animal Damage Control Act*)

dale Refers to a large, wide, open valley.

Darling, J. N. "Ding" "Ding" Darling was an eminent political cartoonist during the first half of this century. With his cartoons, he brought the idea of conservation to the forefront before most people knew what the term **ecology** meant. His interest in the environment did not stop with the pen. He became the chief of the organization that later became the U.S. Fish and Wildlife Service, originated the Federal Duck Stamp Program, which today generates millions of dollars in revenue for wildlife refuges, and then founded the **National Wildlife Federation.**

The J. N. "Ding" Darling Foundation was created in 1962, after his death, by colleagues and friends, to continue his efforts in conservation education. The foundation funds educational programs at the elementary level and provides grants at the university level, all focused on conservation communications and education. The foundation has no paid staff, so every dollar contributed is used for its programs. Write to 881 Ocean Drive, #17E, Key Biscayne, FL 33149.

DDT DDT is a **chlorinated hydrocarbon** type of **insecticide.** First synthesized in the early 1940s, it was considered by many a panacea for pest control. DDT was found to be deadly to insects but appeared to cause little harm to humans. DDT was used to control mosquitoes that carried malaria and sleeping sickness, and is attributed with saving about five million lives during that time. By the early '50s, however, mosquitoes had become resistant to DDT, so it was no longer effective.

DDT, along with other chlorinated hydrocarbons, was later found to harm more than just the targeted insect pests. These pesticides are very persistent, meaning they remain in an ecosystem for long periods, allowing organisms to ingest and accumulate the pesticide within their bodies. DDT became the symbol of **pesticide dangers** in the early 1960s when traces of it (in the form of DDE) were found in many kinds of animals, including humans, due to **bioaccumulation** and **biological amplification.**

During the 1960s, DDT was responsible for the dramatic decline in the number of predatory birds such as ospreys, cormorants, and bald eagles. These populations declined because the insecticide reduced the amount of calcium available for egg production, resulting in thin egg shells that broke.

DDT was banned in the U.S. in 1972; this was followed by an increase in the populations of affected birds. Recent declines have begun once again, however. This is thought to be due to the DDT used in Latin America, where it remains legal, or possibly due to *pesticide regulations* that allow substantial quantities of DDT to be present as an impurity in other pesticides. (See *Carson, Rachel*)

de-icing roads   Spreading road salt is the standard method of melting ice on frozen highways. The salt often damages or kills trees and vegetation, and contaminates groundwater. It has been suggested that one ton of road salt causes about 600 dollars of environmental damage.

Alternatives to road salt include an environmental product developed by Chevron and sold by Cryotech, called Cryotech CMA, made of limestone and vinegar. Cryotech CMA (formerly called ICE-B-GON) lasts longer and causes no harm to plants, but costs about 20 times more than road salt. With competition, environmental awareness, and more acceptance of road salt alternatives, the price of this and similar products will, hopefully, become more competitive. Call (800) 346-7237.

deadtime   Deadtime refers to the amount of time it takes an item to decompose (biodegrade) into useable organic materials that future generations of plants can use. The shorter the deadtime, the more biodegradable an item is said to be. The deadtime of an item differs greatly depending on whether it is in a *landfill,* where water and air are scarce, or exposed to the elements.

For example, most paper products break down quickly under normal conditions, but have a deadtime of many decades in a landfill. Most organic yard wastes, such as grass cuttings, become useable *organic fertilizer,* if left on the lawn, within a few days, but have a deadtime of years in a landfill. Most plastics have a deadtime of hundreds or thousands of years whether they be in or out of a landfill. (See *Grass cycling, Paper recycling, Plastic recycling*)

debt-for-nature swaps   Beginning in 1987, countries experiencing *rain forest destruction* and a large national debt have received interesting proposals, called debt-for-nature swaps. Some organizations have arranged to pay a portion of the country's debt in return for their rain forest conservation efforts. About $60 million has changed hands in debt-for-nature swaps in the past few years. The countries involved have included Costa Rica, Madagascar, the Philippines, Bolivia, and Ecuador. The two organizations most often involved in these swaps are the World Wildlife Fund and **The Nature Conservancy.** (See *Tropical rain forests, Deforestation*)

deciduous Refers to plants that lose their leaves during cold weather or other environmental change such as drought. (See *Evergreen, Temperate deciduous forest*)

decommissioning nuclear reactors Most **nuclear reactors** have a life expectancy of under 40 years, but many are retired from service much earlier. Since these facilities are contaminated with radioactivity, they cannot simply be knocked down and bulldozed under like other buildings. Three options are currently available. The plant can be dismantled and all the radioactive components shipped to **nuclear waste disposal** sites. This produces large amounts of radioactive waste materials and puts workers at risk of exposure to the radiation.

Another option is to mothball the plant until the **radiation** levels drop, before dismantling the plant. This means the plant must be secured for roughly 10 to 100 years before taking it apart and disposing of the radioactive materials.

The third option is to create a solid concrete tomb over the entire reactor to prevent radiation from escaping until the radiation level is insignificant. The tomb would have to last thousands of years before becoming safe, or until a new technology could be developed to decontaminate it. Even if the tomb prevented radiation from leaking into the air, there is still the danger of it leaching through the soil into the water supply.

Japan, France, and Canada plan to mothball their nuclear plants before dismantling them, while the U.S. plans to dismantle most of its plants immediately. The costs for decommissioning these plants are estimated to be between 50 million and a few billion dollars. A small U.S. **nuclear power plant,** decommissioned by the Department of Energy in 1974, took three years to dismantle at a cost of over $6 million. Over 3,000 cubic meters of dismantled materials from the plant were buried.

Since decommissioning costs are so high, about 20 plants worldwide have been shut down but are still waiting to be decommissioned. Many more are approaching retirement age. (See *Nuclear reactor safety and problems*)

decomposer Organisms can be categorized according to how they obtain food to survive. **Producers** (green plants) make their own food, while **consumers** eat the plants or other animals. Decomposers, however, obtain nourishment by eating dead and decaying organisms, called **detritus.** Decomposers include many insects, worms, bacteria, and fungi (molds and mushrooms). They reduce the detritus to the basic organic nutrients so they can be reabsorbed by a new generation of plants and keep the environment free from the remains of all preceding generations of life on earth. (See *Biogeochemical cycles, Food web*)

deep ecology Deep ecologists believe humankind to be the primary threat to our planet's survival. In the book *Deep Ecology: Living Nature as If Nature Mattered,* by Devall and Sessions, the following deep ecology principles are proposed: 1) Humans have no right to reduce the richness of life except to satisfy basic needs. 2) The quality of human life and culture is linked to a substantial decrease in the human population. 3) A decrease in the human population is required, if nonhuman life is to flourish. (See *Ethics, environmental; Philosophers, environmental; Chief Seattle*)

deep-well injection sites **Hazardous wastes** have been disposed for decades by being injected into deep wells in the earth's crust. These wells can be anywhere from 20 to several thousand feet deep. Many of these wells have polluted **aquifers** that supply **domestic water** to millions of homes. (See *Hazardous waste disposal*)

deforestation In 1988, satellite images began to reveal to the public a new threat to our planet. **Tropical rain forests** in many parts of the world are being cut down (with much of the vegetation burned) at a rate that could change the planet as we know it. This destruction has been coined deforestation.

Forests are destroyed to make room for crops and for grazing animals such as cattle, as well as for timber. Deforestation has a significant impact on global environmental problems such as **air pollution** and **global warming.** In addition, the loss of these habitats is forcing many species to **extinction** and dramatically reducing our planet's **biodiversity.** Tropical rain forests are home to half the species of the world, but 40 to 50 million acres of forests are being destroyed annually. That's about 80 acres per minute. In 1950, 30 percent of the earth's land was covered in these forests but today, only 7 percent remains.

These forests are often burned to add the nutrients locked up in the trees back into the soil to act as fertilizer for crops. This large-scale burning releases vast quantities of carbon dioxide, a *greenhouse gas.* It is estimated that one to two billion tons of carbon (in the form of carbon dioxide) are released into the atmosphere annually due to deforestation.

Most deforestation of tropical forests occurs in South America, but ten tropical forests across the globe have been identified by experts as "hot spots." Hot spots are regions being deforested faster than the world average and have large numbers of plant and animal species found exclusively in these habitats. They are: 1) Madagascar; 2) Atlantic Forest of Brazil; 3) Western Ecuador; 4) Colombian Choco; 5) West Amazonian Uplands; 6) Eastern Himalayas; 7) Peninsular Malaysia; 8) Northern Borneo; 9) the Philippines; and 10) New Caledonia. (See *Slash-and-burn agriculture, Clear-cutting*)

dell A small, wooded valley.

demographic transition Demographers who studied human populations of western European countries during the nineteenth century found what appears to be a link between the standard of living and population growth. They proposed the theory of demographic transition. The transition refers to stages these countries went through, and other countries might go through, as they develop in the future.

In the first stage (preindustrial stage) there are high death rates due to harsh living conditions and high birthrates to compensate, hence a stable population. As the country becomes more industrialized (the transitional stage), the death rate lowers due to somewhat improved conditions, but the birthrate remains high, resulting in an increase in population. Most of the existing *less developed countries* are at this stage now.

In the next stage (industrial stage), further technological advances result in lower birthrates due to reasons such as each individual's desire to advance with the society and the belief that children will hinder them in doing so. The birthrate begins to approach death rates. The majority of *more developed countries* currently fall in this stage.

In the last stage (postindustrial) the birthrates reaches the same level as the death rate, resulting in *zero population growth* and finally a reduction in population. This has already occurred in a few countries, such as Germany.

Many people do not believe what has happened in the past in some countries will necessarily hold true in the future for other countries. Rapid and long-term economic growth is necessary if less developed countries are to make the demographic transition. (See *Age distribution in human population, Doubling-time in human populations, Overpopulation*)

demography Demography is the study of *populations* and the factors that make populations increase or decrease in size.

dendrochore Refers to that part of the earth's surface covered by trees. (See *Biomes*)

denizen Refers to a species of plant that grows wild but is thought to have been originally introduced by humans for cultivation purposes. (See *Exotic species, Indigenous, Endemic*)

descriptive ecology Descriptive ecology concentrates on describing the variety of environments found on our planet and the components of each. This was the first approach used to study *ecology* and was very popular in the first half of the 1900s. It is still an integral part of modern-day ecology. Newer approaches include experimental and *theoretical ecology.*

desert Deserts are one of several kinds of *biomes.* The primary factors that differentiate biomes are temperature and precipitation. Deserts receive less than 25 centimeters (10 inches) of precipitation per year, and therefore are inhospitable to many forms of life. The rain is erratic but when it comes, it is intense, so most of the water runs off. Bursts of life follow these heavy but rare rains. Many species of plants are adapted to complete their entire life cycle and produce seeds within a few days, before the water dries.

Although all deserts have low precipitation, different temperature ranges delineate three types. Tropical deserts (like the Sahara) are hot all year; temperate deserts (like the Mojave) are hot in summer and cool in winter; cold deserts (like the Gobi) are warm or hot in the summer and cold in winter. All deserts experience wide temperature differences between day and night.

The relative number of species in a desert is low compared with other biomes, but there is still considerable diversity. Organisms living in the desert possess specialized adaptations to survive. Plants have small or no leaves to reduce surface area, which in turn reduces water loss, while others store water in their fleshy tissues.

There are many kinds of animals in the desert, but they are few in number. They include insects, snakes, lizards, some **grazing** mammals and a few **carnivores.** Birds are common in many deserts. Most desert animals obtain water via the food they eat. They have a waterproof skin or (in the case of insects) waxy cuticle, and many live underground to avoid the day's heat.

Due to the severe conditions, most plants grow very slowly, resulting in fragile ecosystems. Damage by human activities, such as that caused by off-road vehicles, takes decades to recover.

Desert Protective Council, Inc.   The **Desert** Protective Council was established in 1954. Its purpose is to protect desert plants and animals, cliffs and canyons, dry lakes and sand dunes, and historic and cultural sites. Due to the increase in destruction of the desert and its resources, support for this council has increased considerably in recent years. Write to P.O. Box 4294, Palm Springs, CA 92263.

desertification   When rangeland is overgrazed or cropland is overcultivated, soil nutrients are lost and **soil erosion** is likely. This results in the land becoming more like a **desert** ecosystem than it once was. If the land becomes at least 10 percent less agriculturally productive than it once was, the process is called desertification. "Severe desertification" means the productivity of the land has been reduced over 50 percent. Desertification can occur naturally along the fringes of existing deserts, due to drought or climate change.

Desertification can be controlled, in part, with proper land management practices that prevent soil erosion. **Sustainable agriculture** practices are designed to prevent desertification . (See *Soil conservation*)

detritus   Refers to partially decomposed organic matter. (See *Decomposer, Food web, Leaf litter*)

diatoms   Refers to a type of algae with oddly shaped cell walls containing silica. Diatoms make up a large portion of most **plankton.** (See *Aquatic ecosystems*)

dientomophilous Refers to a species of plant that has two different types of flowers for the purpose of attracting two different kinds of insects to pollinate the plant.

dioxin Dioxin is a toxic chemical by-product of many industrial processes. It forms when chlorine is exposed to extreme heat. This **toxic waste** is believed to cause cancer and birth defects and to affect the human nervous system. It is found in industrial waste **effluent** entering the water and is especially prevalent in waste from pulp and paper mills. Dioxin is found in soils, and traces have even been found in mothers' **breast milk.** (See *Bioaccumulation*)

directive coloration Refers to markings on an organism that diverts the attack of a predator to nonvital body parts. For example, eye spots on a butterfly's wings draws the attention of a bird away from the insect's body. The butterfly can survive a bite to its wing far better than one to its body. (See *Mimicry, Aposematic coloration*)

dishwasher "rinse hold" You can conserve about 5 gallons of water each wash by not using the rinse hold cycle. (See *Tub bath vs. shower, Domestic water conservation*)

disposable diapers Over 10 billion disposable diapers are used in the U.S. annually. It is estimated that roughly 4 percent of U.S. **municipal solid waste** consists of disposable diapers, which translates into a cost of around $100 million in waste management costs annually.

Cloth diapers, when discarded, decompose within months, while disposable diapers (which contain plastic) take years. Estimates for decomposition range from decades to several hundred years. "Biodegradable" disposable diapers are claimed to break down quicker, but at present appear to be more a **green marketing** gimmick than a technological advance. In an effort to make disposable diapers truly biodegradable, the industry is spending large sums of money in research and development of new types of disposable diapers and in **composting** facilities that could lead to real biodegradability. (See *Landfills, Plastic recycling, Green products*)

disposal fee These fees help pay for the disposal costs of certain products. They are usually charged to the consumer, but can also be charged to the producer (who will probably pass the expense back to the consumer). These fees can be "front-end" fees, charged when the product is purchased or "back-end" fees, charged when the product is disposed.

Back-end fees include entrance or dumping charges at **landfills** or **incineration** facilities. Front-end fees include taxes or deposits on products such as car batteries and tires, or beverage bottles. For example, Connecticut and Washington charge a five-dollar deposit upon purchase of a car battery, and Maine charges a five-dollar tax on major appliance purchases. (See *Bottle bills, Green tax, Appliance recycling, Recycling*)

DL50   A measurement of "drought lethality," which is the amount of "dryness" that injures 50 percent of a population. (See *Limiting factor, Tolerance range*)

dolphin safe tuna   See *Driftnets, Flipper Seal of Approval.*

domestic water pollution   After **domestic water** has been used, it returns to the environment. If released into a septic system, it may contaminate the **groundwater.** If it enters a **sewage treatment** system, sooner or later it reaches streams, rivers, and the ocean. Domestic waste water includes large amounts of organic matter (leftover food and human waste) that act as food for numerous microbes. High concentrations of these microbes affect **aquatic ecosystems** by depleting oxygen from the water, destroying the aquatic **food chain.**

Domestic waste water also includes soaps and detergents, some of which contain chemicals that affect aquatic ecosystems. For example, many detergents contain phosphates, which is a **nutrient.** When the waste water enters a pond or lake, increased concentrations of phosphates produces a population explosion of algae that fouls the water, resulting in **eutrophication** and the death of the ecosystem. For this reason, many states have banned the use of phosphates in detergent and have improved their waste water treatment procedures to remove these substances.

domestic water use   Domestic use of water is one of four forms of **water use.** Over 25 billion gallons of water are used each and every day in the U.S. for domestic water. Most domestic water comes from underground **aquifers.** A typical family of four in the U.S. uses over 100 gallons of water each day for flushing toilets and another 100 for watering the lawn. About 75 gallons are used each day for bathing and 30 for laundry. Another 18 are used each day for dishes, and about 6 for drinking and cooking. (See *Drinking* *water, Domestic water pollution, Domestic water conservation)*

domestic water conservation   Any steps taken to reduce the use of **domestic water** is considered domestic water conservation. This includes stopping leaky faucets and valves, using low or ultra-low flush toilets, high-efficiency shower heads, and faucet aerators. Using these devices can reduce the amount of water used in the home by more than 30 percent. (See *Tub bath vs. shower, Domestic water use, Domestic water pollution, Domestic water conservation)*

dominants   Some **ecosystems** have a single species that is so abundant, that it dictates the overall characteristics of the area. For example, the sugar maple is the dominant species in the Eastern forests and exerts considerable control over what other types of organisms can survive in the region. (See *Biomes, Keystone species, Exotic species, Indigenous, Endemic)*

doubling-time in human populations   The world's people **population** is growing at around 1.8 percent per year, which seems deceivingly small. Compounding this increase, however, actually results in rapid growth. For example, if you took 100 percent of a population and increased it by 10 percent the first year, you now have a population 110 percent the original size. A 10 percent increase the next year means a 10 percent increase of 110 percent, which results in a population 121 percent of the original size. The deception becomes clear after only a few years of compounded growth.

One of the easiest methods of understanding the magnitude of this increase is to look at how long it takes for an existing population to double. Countries that have growth rates of 4.0 percent or higher, such as Kenya and Saudi Arabia, double their population every 17 years or less. Growth rates from 3.0 to 3.99 percent, such as occur in Nicaragua, Algeria, and Iran, double every 18 to 23 years. Growth rates between 2.0 and 2.99 percent, such as those of India and Mexico, double every 24 to 34 years. Growth rates between 1.0 and 1.99 percent, which countries such as China, Australia, and Argentina have, double every 35 to 70 years. Growth rates between 0 and .99 percent, such as are common in the U.S., Canada, Russia, and most of Europe, double every 71 years or more while a few countries, such as Sweden and Germany, have declining populations. (See *Carrying capacity, Overpopulation)*

Douglas fir   The Douglas fir is a classic **old-growth forest** tree. They are cherished by environmentalists and prized by foresters. A typical tree lives 400 to 1,000 years, grows 300 feet tall and 50 feet around. These trees are found throughout the Rocky Mountains, from the Pacific Northwest and down to western Texas. A live tree removes over 400 tons of carbon from the air during its lifetime and provides a habitat for hundreds of animals, small and large. Foresters are more interested with the fact that a single tree provides enough wood to build a single-family home. (See *Ancient forest, Coast Redwood, Forest Service*)

Douglas, Marjory Stoneman   Marjory Stoneman Douglas, a 101-year-old writer and reporter, is the founder and first president of Friends of the Everglades. This support group was established to prevent the development of the everglades. Many people believed that the everglades (along with other **wetland** habitats) was nothing more than a snake-infested swamp. Ms. Douglas had other ideas, which she researched and wrote in a book that was published in 1947, *The Everglades: River of Grass*. This woman and people like her are at the vanguard of protecting important but fragile **ecosystems** such as the everglades and other wetlands. (See *Environmentalist, Environmental organizations, Estuary and coastal wetlands destruction*)

dredging   Dredging is the mechanical removal of **sediment** that builds up at the bottom of a body of water. Once dredged, the sediment is disposed of. In industrial regions, this sediment often contains contaminated wastes. Dredging these bodies of water poses environmental problems in two forms. First, during the dredging process, much of the contaminated sediment becomes suspended and contaminates the **aquatic ecosystem** all over again. Second, the contaminated sediment must be disposed of somewhere and becomes part of the **NIMBY** (not in my backyard) syndrome. (See *Water pollution, Toxic wastes, Heavy-metal pollution*)

driftnets   Driftnets are sophisticated fishing devices used to catch large volumes of tuna, salmon, and squid in a relatively short period of time. In addition to the intended catch, the nets entangle large quantities of other types of fish and marine life. This unintentional catch is called **bycatch.**
   Driftnets are finely woven nylon-mesh nets up to 50 miles in

length that drop into the deep waters about 40 feet. As unsuspecting fish swim into these huge walls of mesh, they become entangled and soon die. It is estimated that 2,500 ships put out over 50,000 miles of driftnets and their coastal water equivalent, *gillnets,* every night. Hundreds of miles of these nets become entangled or lost at sea each year where they continue to trap countless fish and marine mammals.

These nets not only decimate entire populations of targeted fish, but also wipe out tens of thousands of bycatch species, including dolphins, sharks, whales, marine turtles, and sea birds. The killing of these nontargeted species in vast numbers severely damages many *marine ecosystems.*

The U.S. Marine Mammal Protection Act protects whales, dolphins, and some other marine life, but excludes "incidental" catches, which includes the bycatch of driftnets. The Fisheries Conservation and Management Act does not address driftnets either. Public pressure within the U.S. has resulted in many tuna canners selling dolphin safe tuna, which assures the consumer that the tuna was not caught in driftnets, or *purse seine nets.* Internationally, Japan, Taiwan, and South Korea have the largest fleets of driftnet ships and have resisted any attempts at restricting the practice. (See *Turtle excluder device*)

drinking water   In the majority of *less developed countries* throughout the world, safe drinking water is rare and waterborne diseases common. Even many of the more developed countries are affected by this problem.

In the U.S., about half of the drinking water comes from *groundwater* (wells) and the other half from *surface waters* (lakes, rivers, etc.). Hundreds of communities have reported outbreaks of waterborne illnesses and diseases in recent years. Some are caused by high levels of bacteria and other microbes, but others are caused by contaminants, such as *pesticides, heavy metals,* radioactivity, gasoline additives, and cleaning solvents, turning up in water supplies. These contaminants are polluting both the groundwater and surface waters.

A few examples of drinking-water problems include the following. Studies have found that one in six people in the U.S. drinks water containing excessive amounts of lead. During the growing season, half the streams and rivers in the Corn Belt contain unhealthy concentrations of pesticides, and one in three cases of gastrointestinal illness are caused by microbes in drinking water.

Private well water must be tested individually, but the EPA is responsible for setting guidelines for municipal water supplies. The majority of contaminants found in water, however, are not regulated by the EPA or any government agency. Many states have created stricter and more comprehensive regulations to fill in for the EPA.

drip irrigation   Most types of *irrigation* methods lose large volumes of water by evaporation and by seepage. In the 1950s, Israel developed the drip method, in which perforated pipes are installed just below the soil surface in the root zone of the plants. As the water drips out, it is delivered directly to the plant roots, reducing the amount of water loss from the usual 50 percent to less than 25 percent. This method is used on only a small fraction of the irrigated lands in the U.S. (See *Monoculture*)

duff   When *leaf litter* (dead leaves, pieces of bark, and twigs that lie on the forest floor) breaks down into fine, powdery substance, it is called duff. (See *Decompose*)

dulosis   A relationship in which worker ants of one species capture the brood of another and rear them as slaves. (See *Symbiotic relationship, Competition*)

dust bowl   Between 1933 and 1939, about 150,000 square miles of farmland in the Great Plains of the U.S. was denuded of topsoil. Huge dust storms turned day into night. The land had been overplowed and overplanted for decades, which turned the once fertile soil to powder. On May 11, 1934, one storm is believed to have blown away 300 million tons of topsoil. In many of the affected states, 60 percent of the entire population was forced to move to other locations in search of a livelihood.

The causes of *soil erosion* during the Great Dust Bowl are understood and can be significantly controlled by *soil conservation* measures including *conservation tillage, contour farming, terracing, strip cropping,* and *windbreaks,* among others. Poor farming practices however, still exist on many farms in the U.S. and especially in many *less developed countries,* resulting in devastating losses of topsoil on a global basis.

dust dome   Heat produced and absorbed by cities forms an *urban heat island.* Under certain weather conditions, this heat island generates its own air currents, acting like a microclimate above the city. The circulating air in this heat island traps pollutants and dust particles, forming a dust dome or bubble over the city, which increases levels of *air pollution.* The dome is usually blown away when a strong cold front moves into the area.

E Magazine   Called *The Environmental Magazine* by the publisher, this is an entertaining, easy-to-read **eco-magazine** available on the newsstand. Anyone with even a remote interest in our environment would enjoy it. Published bimonthly, it contains lots of interesting tidbits plus many informative full-length articles. Call (800) 825-0061.

E-lamp   See *Radio-wave lightbulb.*

Earth Communications Office (ECO)   ECO is a nonprofit, nonpartisan organization of actors, movie directors, producers, and writers dedicated to making the public aware of environmental concerns. Since celebrities readily receive attention, they use TV shows, films, tapes, and magazines to heighten the public's awareness of environmental issues. They believe the entertainment industry can greatly influence the public by conveying these messages through public service announcements as well as incorporating environmental issues into their story lines. Write to 1925 Century Park East, Suite 2300, Los Angeles, CA 90067. (See *Environmental Media Association, Environmental Film Association*)

Earth Day   The first Earth Day was the brainchild of Gaylord Nelson (senator from Wisconsin at that time) and was held on April 22, 1970, across the U.S. Many people credit the first Earth Day as the unofficial beginning of the modern-day environmental movement. Millions of individuals, businesses, and government agencies participated in environmental education and activism. Polls indicated a dramatic increase in environmental awareness following the event, which has been held every year since. By 1990, the 20th anniversary of the first Earth Day, it had evolved into a worldwide event for environmental awareness and education. Earth Day Resources can help you organize an Earth Day event. Its address is 116 New Montgomery Street, #530, San Francisco, CA 94105. Call (800) 727-8619.

Earth Island Institute   This organization has created and manages about 20 innovative projects, each focusing on a major environmental concern. Some of these projects are: The Climate Protection Institute, The Fate of the Earth Conferences, The Environmental Litigation Fund, Friends of the **Ancient Forest,** The **Rainforest** Health Alliance, The Sea Turtle Restoration Project, and the **Urban** Habitat Program. All the projects are education oriented. Since its inception in 1982, the institute has established a respected worldwide network of environmental leaders. It has 32,-000 members and the annual membership fee is 25 dollars. Write to 300 Broadway, Suite 28, San Francisco, CA 94133.

earth jazz   Earth jazz, also called earth music, "celebrates the cultures and the creatures of the whole earth," according to one of the best-known earth-jazz musicians, Paul Winter. Animal sounds become part of the music such as in "Wolf Eyes," which is a duet with a timber wolf.

Earth Summit   During the first 12 days of June 1992, world leaders gathered to discuss the planet's environmental problems in Rio de Janeiro. This worldwide conference was called Earth Summit—the **United Nations Conference on Environment and Development (UNCED).** Along with political leaders, more than 30,000 activists, governmental and nongovernmental delegates, religious leaders, and corporate CEOs also attended this conference. The foundation of this conference was based on five documents: **The Treaty on Biological Diversity,** The **Convention on Climate Change, Forestry Principles,** The **Rio Declaration** and **Agenda 21.**

Although much was accomplished during this conference, many significant issues were omitted. This was mainly due to a consensus rule, mandating that all issues must be approved by all 172 countries. Overall, the result of the Earth Summit did produce a new agenda for protecting the environment and laid the foundation for future progress. It also proved that participants from every country, culture, religion, and nationality can come together and unite for one common goal, to protect our environment.

earthworms   Earthworms are vitally important in conditioning **soil** for plant growth. In their quest for food, they eat through soil, mixing inorganic with organic material. Their tunneling improves the soil's ability to absorb air and drain water. Two and a half acres of soil may contain 500,000 earthworms that will eat through 10 tons of soil in a year, producing a crumbly textured soil, well conditioned for plant growth. (See *Soil organisms, Soil texture*)

eclosion   Refers to an adult insect emerging from a pupal case. For example, a butterfly emerging from its cocoon. (See *Insect*)

eco-   "Eco" is derived from the Greek word for house. The prefix has become the buzzword, or "buzzprefix," for anything pertaining to our environment. There is ***ecotravel, eco-terrorism,*** and ***ecopreneurs,*** as well as ***eco-magazines,*** eco-movies, and eco-books. "Eco" is also often used in many ***green marketing*** campaigns to make the consumer believe the manufacturer is "environmentally concerned". (See *Green products, Environmentalist*)

eco-conservative   Refers to a person who is politically cautious about making changes that might harm the ***biosphere.*** An eco-conservative allows change to occur only if there is a scientific basis that the change will not cause harm to the environment. Eco-conservatives allow or disallow these changes to occur by voting for or against politicians and referendums at the local, state, and federal levels. (See *League of Conservation Voters, Greenscam, Environmentalist, Environmental literacy*)

eco-magazines   Magazines that cover environmental issues have been around a long time. A new surge in environmental awareness over the past few years, however, has carried in a new bevy of eco-magazines and helped expand a few old standards. ***E Magazine, Buzzworm,*** and ***Garbage*** are three excellent newcomers written for the new wave of environmentalists.

Colorful, informative magazines like *Sierra* and *Audubon* have been popular for years among their members. In an effort to compete more aggressively with the newcomers, they are now available on newsstands. ***Greenpeace*** has an excellent magazine with hard-hitting investigative reporting, and *Worldwatch* is always packed with vital information about global issues. Most large bookstores and newsstands carry many of these magazines.

eco-terrorism   Refers to aggressive or violent acts designed to either a) protect the environment; or, b) use the environment as a pawn during a dispute. Trying to stop ***deforestation*** by driving spikes into trees, endangering the lives of loggers using chain saws, would be an example of the former. Using environmental degradation as a weapon, such as occurred during the ***Persian Gulf War,*** is an example of the latter. (See *Ethics, environmental*)

ecocity   About 75 percent of all people living in industrialized countries and 50 percent of those in the *less developed nations* live in cities. Proposed cities that work with the environment, instead of against it, are referred to as ecocities.

Most ecologists that study **urbanization** don't think there is anything inherently wrong with cities, but that something is wrong with the way they are currently designed and built. Ecocities emphasize sustainability by conserving land and other resources and by polluting less. One of the primary offenders in cities is the need for commutation and the **fossil fuel**–burning automobile that fulfills the need. Ecocities would be designed to contain relatively dense populations to reduce the need to get from here to there. Walking and biking should be the normal form of transportation to and from work or for shopping. **Mass transit** would be the only acceptable alternative. This means homes, business, and shopping areas must be close to one another. They would also be near resources such as water and farms so it would not have to be transported in. (See *Urban sprawl, Bicycles, Automobile fuel alternatives, Conservation districts*)

ecolinking   Refers to the use of computer technology (communications hardware and software, global networks, electronic mail, bulletin boards, and online services) by people in all parts of the world, to share thoughts, ideas, and research on environmental issues. (See *Econet, MNS Online, Computer networks, Computer software*)

ecological studies   Ecology can be studied from many different angles. It can be studied according to the **habitat** of interest (terrestrial, marine, and freshwater), by the types of organisms of interest (plants, animals, microbes), by the **level of biological organization** (from a single organism to an entire ecosystem), or by the methodology used during research. All of these approaches are usually used in conjunction with one another during ecological research. (See *Ecological studies methods*)

ecological study methods   *Ecology* can be studied using one of three different types of approaches or methodologies: **descriptive ecology, experimental ecology,** or **theoretical ecology**. Most early ecological research was descriptive (qualitative) in nature, but today the field is primarily experimental and theoretical (quantitative). (See *Computer modeling*)

ecology   Ecology is the study of the relationships that exist between all the components of an **environment,** including the interactions between organisms with other organisms and with the nonliving components of the environment, such as the geography and climate of a region. You can think of the environment as a set of dominoes, and ecology as a study of the domino effect. (See *Environmental science*)

EcoNet   EcoNet is an environmental **computer network.** Founded in 1987, it provides worldwide access to information about environmental issues. Members can contact other members in over 70 countries via electronic mail. For information, contact The Institute for Global Communications, 18 De Boom Street, San Francisco, CA 94107; (415) 442-0220. (See *Computer software, Environmental*)

economic vs. sustainable development   Unfortunately, economic development is often accompanied by a decline in the quality of our environment. Until recently, there was little concern for any negative environmental impact, as long as the "progress" was good for the economy. Today many people are trying to "figure-in" to the economic picture the impact to our planet. Removing a forest is good for the economy since it provides wood and other materials, but if the land erodes away and becomes unproductive for future generations, was it truly progress?

A new view of our economy is based on sustainable development, which takes any negative impact on the environment into consideration when assessing economic growth. Sustainable development has been defined as addressing present needs, without compromising future needs. Instead of using economic health indicators such as the gross national product, new ways such as the **Index of Sustainable Economic Welfare (ISEW),** which tries to balance the economic worth of a product or service against the environmental loss, are used. (See *Economy vs. environment, GATT, President's Council on Competitiveness*)

economy vs. environment   Many people are under the false assumption that whatever is good for the environment must be bad for the economy. Concerns about the quality of our environment have created their own booming global business. Professions in environmental sciences, especially environmental cleanup, are the fastest growing sector in the U.S. economy, even outpacing computer technologies. About three million people are employed either directly or indirectly in environmental cleanup work in the U.S., with over 60,000 companies participating. New technologies in environmental sciences are helping fill the decline in the aerospace and defense industries, and will continue to expand rapidly, producing many more jobs. During the 1990s, about 3.5 trillion dollars will be spent worldwide on environmental cleanup. The country best prepared to fill this technological need will benefit, economically, the most.

The problem is not the environment versus the economy; instead it is the age-old problem of change. Individuals who earned a living tending horses or building carriages probably believed the automobile was bad for the economy. Environmental protection appears to be an enormous opportunity, not an impediment. (See *Economic vs. sustainable development*)

ecopreneur   Refers to an entrepreneur who specializes in environmental products or services. (See *In business*)

ecosystem   Refers to a description of all the components of a specified area, including the living (organisms) and the nonliving factors (such as the air, soil, and water), and the interactions that exist between all these components. These interactions result in a relatively stable assortment of organisms and involve the continuous cycling of nutrients between the components.

The area defined as an ecosystem is arbitrary. It can be a complex biological system, such as a **biome** or a **habitat** such as a lake or forest. However, smaller entities, such as a rotting log, can be considered and studied as ecosystems. (See *Ecology, Food webs, Food pyramids, Habitats, Succession*)

ecotourism   Ecotourism refers to the business of nature travel (**ecotravel**). Areas that have a unique natural attraction often center their economy around those people willing to pay to see or participate in these attractions or activities. Some countries' entire economy revolves around ecotourism, such as in the Caribbean and Belize in Central America. Increased revenues are often accompanied by increased environmental destruction and pollution. Regions that are big on ecotourism must rigorously manage their natural resources to prevent the tourists they seek from destroying the attractions they come to see. (See *Ecopreneur, In business*)

ecotravel   Refers to travel in which the main focus is environmental awareness. Activities range from purely educational, such as studying ecosystems or indigenous peoples; hobby oriented, such as photo expeditions into exotic habitats; or thrill-seeking, such as shooting the rapids or mountain climbing. There are over 100 private organizations in the U.S. that specialize in ecotravel adventures. Travel agencies often have special ecotravel guides. For more information, call the **EcoTourism** Society at (703) 549-8979. (See *Ecopreneur, In business*)

ectoparasite   A **parasite** that lives outside of its host, such as a tick. (See *Parasitism*)

edaphology   The study of **soils** with a concentration on the use of the land for cultivation. (See *Monoculture, Irrigation*)

effluent   The discharge and flow of liquid waste into the environment. (See *Sewage, Industrial water pollution*)

Ehrlich, Paul R.   Paul Ehrlich is a population biologist and ecologist who has become the authority on the effects of **overpopulation** on the environment. Professor Ehrlich's 1968 book, *The Population Bomb,* enlightened the public about the forthcoming population crisis. He continues to educate about this problem with more recent books, including *The Population Explosion,* published in 1990. (See *Carrying capacity, Populations, Doubling-time in human populations, Competition*)

electromagnetic radiation   Anything connected to an electrical circuit radiates both an electric field and a magnetic field. Some research suggests that some magnetic fields might pose health risks. Magnetic fields created by electric circuits are similar to those created by common household magnets. The magnetic fields of both pass through most materials, cannot be sensed by humans, and become weaker the farther they are from the source. This, however, is where the similarities between household magnet fields and electric circuit magnetic fields end.

The magnet produces a field that has a constant strength, but the electric circuit field reverses itself at a frequency of 60 times a second (60 hertz), which is standard in the U.S. and Canada. A field that fluctuates at this frequency is in the ELF (extremely-low-frequency) range, which is the specific type of electromagnetic radiation that may cause health problems. (See *Extremely-low-frequency magnetic radiation, Indoor radiation*)

eleventh commandment   Individuals involved in the population control movement have added an informal eleventh commandment that states, "Thou shalt not transgress the **carrying capacity** of the environment." (See *Overpopulation, Doubling-time in human populations, Population*)

elfin forest   A forest located at elevations higher than normal, characterized by stunted tree growth. (See *Limiting factor, Zones of life*)

emergent plants   Refers to aquatic plants that are rooted at the bottom but protrude through the water's surface, often to flower. Water lilies, cattails, and bullrushes are all emergent plants. (See *Standing water habitats, Aquatic ecosystems*)

emigration   Emigration is one of two forms of international **migration.** Emigration refers to the movement of people out of their native country into other countries. **Immigration** is the other form of international migration when the movement is into a country. Both forms of migration, along with **fertility** and **mortality** dictate changes in the size of a **population.**

endangered species   An endangered species refers to organisms that exists in small enough numbers that they might become extinct unless certain factors are changed immediately. In other words, some positive form of human intervention is needed to save them. There are about 300 species currently on the endangered species list, which is controlled by the Office of Endangered Species (part of the Department of the Interior). The black rhinoceros, black-footed ferret, California condor, Manatee, and Mission blue butterfly are all examples of endangered species. (See *Endangered Species Act, CITES, Northern Spotted Owl, Threatened species*)

Endangered Species Act   The Endangered Species Act became law in 1973. It gave the federal government jurisdiction over the management of any organism listed as an **endangered species.** The original act had strong language and the clout to enforce the words. It stated that no government agency could perform any activity that would lead to the extinction of an organism on the list and that all government agencies must cooperate to prevent extinction.

Since its inception, the Act has been underfunded and gradually watered down. In 1978 an Endangered Species Review Committee was established that could override the Act if the economic advantages outweighed the ecological effects. Also, amendments to the Act make it considerably harder to add species to either the Endangered or Threatened Species lists. As of 1992, over 1,000 species have been placed on a "waiting list" until enough monies are budgeted.

During the Bush administration, a special panel, unofficially called the "God Squad," had the power to overrule the placement of any animal on the endangered species' list, if listing would pose a burden to business interests. It could even remove species already on the list. (See *CITES*)

endemic　Refers to something found only in a certain region. For example, many plants and animals are endemic to areas being deforested, or a disease may be endemic to a certain region. (See *Exotic species, Indigenous*)

endoparasite　A *parasite* that lives inside its host, such as a tapeworm. (See *Parasitoid, Biological control*)

energy (fuels)　Readily available sources of energy (fuels) are intimately related to the development of nations. These fuels have had a dramatic impact on the environment in the past and will in the future. Sources of energy can be divided into two categories: 1) *fossil fuels,* which are *nonrenewable energy* sources, supply over 80 percent of the world's energy demands; and 2) *alternative energy* sources (alternatives to fossil fuels) include many types of *renewable energy* sources, and *nuclear power.* (See *Cool power*)

energy consumption, historical　As civilizations became highly developed, the need for readily available fuels grew dramatically. Even though the world population doubled between 1900 and 1990, we consumed 12 times more energy over the same period. The 12-fold increase was not due to the increase in the number of people as much as a 30-fold increase in the number of products produced during the same period.

About 80 percent of the world's energy supply today comes from *nonrenewable energy* sources. With continued population growth and increased productivity, the demand for energy increases. *Alternative energy* sources will have to be found since nonrenewable sources will be depleted.

Historically, the switch to an alternative energy source (once available), such as coal in the nineteenth century, and oil and gas in the twentieth century, takes about 50 years to be accepted, implemented, and put into large-scale use.

energy plantations   Refers to farms that grow crops specifically for use as a fuel source for **biomass energy**.

energy pyramid   Energy pyramids illustrate how organisms obtain energy and how the total amount of available energy changes as it flows through an **ecosystem**. Each level of the pyramid, called a trophic level, represents a different method of obtaining energy. The bottom level represents **producers** (plants) that capture the sun's energy during photosynthesis, creating their own food (chemical energy). Each subsequent trophic level represents different types of **consumers**. The second level contains primary consumers, which eat the producers. The next level contains secondary consumers, which eat the primary consumers. This is followed by the tertiary consumers and possibly a fourth level at the top of the pyramid.

Organisms in each trophic level use (burn up) most of the energy available to them. They use this energy for breathing, growing, moving, reproducing, making sound, and all of life's other activities and functions. About 90 percent of the chemical energy available at each level is used by that level, leaving only 10 percent for the next higher level.

If we assume the producers (bottom trophic level) begin with 100 percent of the energy entering the pyramid, then the next higher level, primary consumers, have only 10 percent available to them. Secondary consumers are left with only 1 percent, and tertiary consumers and quartenary (fourth) consumers have .1 percent and .01 percent of the original energy left to them, respectively.

With less and less energy available, the organisms at the higher levels have less food available to them, resulting in fewer organisms at each higher level. For example, a field of grass might contain hundreds of field mice, 10 snakes, and only one hawk.

Energy pyramids illustrate why it is far more efficient to feed people plant foods (grains) than animal foods (meats). Assume a field yields 100 pounds of grain. Feeding people from this trophic level provides the entire 100 pounds of food. However, if the same 100 pounds of grain was fed to cattle, most of the energy would be used up (by the grain and cattle), resulting in only a few pounds of meat for people to eat. (See *Food chains, Thermodynamics, second law*)

energy sources, historical   Wood was the first source of fuel. It was used to cook food, heat living areas, and mold metals into utensils, tools, and weapons. As the demand for wood grew, the supply diminished in many parts of the world and new alternatives had to be found. Western Europe began running out of wood in the thirteenth century, but the North American forests kept wood as the primary fuel in the U.S. through the mid 1800s.

As the supply of wood diminished, ***coal,*** a ***fossil fuel,*** became popular. During the early eighteenth century, those countries with large supplies of coal participated in the Industrial Revolution. In 1850, 90 percent of the U.S. energy supply came from wood, but by 1900, 70 percent was supplied by coal.

At the same time coal was taking the spotlight away from wood, ***oil*** wells were being drilled. From the mid 1800s to about 1900, oil was used primarily for lantern kerosene while the gasoline component was considered a waste by-product. In 1870, oil supplied only 1 percent of the U.S. total fuel needs. During the early 1900s, however, the automobile and its thirst for gasoline made our dependence on oil surge. Today the U.S. depends on oil for about 40 percent of its fuel and only 23 percent for coal.

Just as gasoline was originally considered a waste by-product of crude oil, ***natural gas*** used to be burned off at the oil well as waste. Natural gas was ushered in as a good source of energy, especially for heating. During the 1920s, about 5 percent of the U.S. energy supply came from gas, but it has gradually increased and now supplies about 24 percent, even surpassing coal.

Hydroelectric power also came into its own during the early 1900s and currently supplies about 21 percent of U.S. energy, but only about 6 percent of the world's power. ***Nuclear power*** became popular during the later part of the twentieth century, along with many other forms of ***alternative energy,*** all of which are now supplying the U.S. with about 10 percent of its energy needs.

energy use   Energy use can be divided into four major categories: 1) residential and commercial; 2) industrial; 3) transportation; and 4) electrical utilities. In general, less developed countries use a greater percentage of the energy for residential purposes while more developed countries use a greater percentage for industry and transportation.

The energy use of a country and its people can be profiled by looking at the amount of energy used per person for each of the categories mentioned above. For example, the amount of energy used for transportation by each person in the U.S. is 100 gigajoules, compared to 40 in Denmark, 25 in Japan, 12 in Mexico, and two in Zimbabwe. (See *Energy sources, historical; Energy consumption, historical; Joule, Negawatts*)

entomology   The study of *insects.*

Environet   Environet is an electronic (computer) bulletin board, sponsored by **Greenpeace,** for anyone interested in environmental issues. It contains press releases, news bulletins, and other information that can be downloaded. Open electronic forums are also available for ongoing communications on various environmental topics, or simply to chat with others. (See *Computer networks, MNS Online, News Service, Environment; EcoNet*)

environment   The environment refers to all the living and non-living components, and all the factors, such as climate, that exist where an organism lives. The plants and animals, mountains and oceans, temperature and precipitation all make up an organism's environment. Environment assumes the perspective of the organism being studied or discussed (e.g., the rabbit's environment, or dumping waste damages our environment).

This term is often confused with **ecology,** which is also the study of these components and factors, but more important, the relationships that exist between them. Ecology is the study of how the living parts interact with the nonliving parts, and how factors, such as weather, impact all the parts. You can think of the environment as an assortment of dominoes around you, and ecology as a study of the domino effect, or the impact of one domino on the others. (See *Environmental science*)

Environmental and Energy Study Conference
This is the largest legislative service organization in Congress. It services over 335 senators and representatives by providing objective analysis of environmental, energy, and natural resource issues. Government agencies and outside experts are used in their extensive studies. It is chaired by two Democrats and two Republicans. Its findings on these issues have broad ramifications in Congress.

Environmental Careers Organization, Inc., The
Originally called the CEIP Fund. This is a national environmental
careers organization that places applicants into entry-level intern-
ships in environmental fields lasting three to nine months. Stipends
for the interns average 350 dollars per week. College juniors and
seniors, graduate students, recent grads, and people changing their
careers are eligible to apply for a CEIP associate position. The
program is sponsored by major corporations, environmental organizations, and local, state, and federal government agencies. Publishes *The Complete Guide to Environmental Careers*. Write to 286
Congress Street, 3rd floor, Boston, MA 02210.

Environmental Defense Fund (EDF) The EDF was
established in 1967 and has about 150,000 members. It has broad
interests, but specializes in environmental quality and human health
issues. It looks for new and creative ways to resolve long-standing
environmental problems and is at the forefront of legislative action.
The *New York Times* referred to the EDF as "one of the most
powerful environmental organizations in the world." The annual
membership fee is 20 dollars. Write to 257 Park Avenue South, New
York, NY 10010.

environmental education Almost all the major environmental organizations are involved in education, usually having separate education committees, offices, or programs. Some
organizations, however, specialize in environmental education by
acting as a clearinghouse, directing people where and how to find
the educational materials and ideas they need. The **North American Association of Environmental Education** and the Alliance for Environmental Education, at (703) 335-1025, are two
examples. The **Environmental Media Association** offers information on educational films and other media. The **EPA** has an
Office of Environmental Education, which can be reached at (202)
260-4962. (See *Environmental literacy*)

Environmental Film Resource Center The Environmental Film Resource Center is an organization dedicated to collecting environmental films for individuals or corporations, promoting
the distribution of these films at educational institutions, and lobbying for the exhibition of these films on television throughout the
world. They have programmed environment films used for National
Geographic Explorer, The Discovery Channel, and the Disney

Channel, as well as working with ABC's 20/20, gathering important resources for a program on *indoor pollution*. Write to 324 N. Tejon Street, Colorado Springs, CO 80903, or call (719) 578-5549. (See *Environmental Media Association, Environmental education, Earth Communications Office (ECO)*)

environmental literacy   Refers to the basic level of understanding an individual should possess to make intelligent decisions about managing our environment. Why *recycle,* if you don't understand what it accomplishes? Why vote for or against a *landfill* or a *waste-to-energy power plant* about to be built in your town, if you don't understand the advantages and disadvantages of each? Why be concerned if the fruit you eat is labelled with *food irradiation* symbols or the vegetables you eat have been waxed? To be environmentally literate means you are not only interested in our environmental problems, but you understand these problems and therefore can help to resolve them in a responsible way. (See *Environmentalist, Environmental organizations, Environmental education, The League of Conservation Voters, Eco-magazines, Fruit waxing*)

Environmental Media Association   This is a nonprofit, Hollywood-based association dedicated to urging major studios to use environmental themes in television shows and movies. It holds the Environmental Media Awards competition each year, which recognizes excellence in this field. (See *Environmental Film Resource Center, Earth Communications Office*)

environmental organizations   Environmental activism has reached new heights. This can easily be seen by the number of environmental *nongovernmental organizations (NGOs)* that exist. Whatever information you may require or whichever topic or issue you wish to get involved in, there is an organization for you to contact and join. Many of these groups are listed as entries in this book. There are books available that specifically list environmental organizations. One such book is *Your Resource Guide to Environmental Organizations,* published by Smiling Dolphin Press in Irvine, California. A more comprehensive list is the annual National Wildlife Federation's *Conservation Directory.* (See *United Nations Environment Programme*)

Environmental Protection Agency (EPA)

The EPA was established in December of 1970. Its purpose was to consolidate federal environmental activities that were being carried out by numerous agencies. The EPA was authorized to manage air and *water pollution,* and *pesticides.* It was also charged to manage *municipal solid waste, hazardous waste,* and *radioactive waste* disposal. Its headquarters are located in Washington DC, and there are 10 regional offices across the country. (See *Environmental Protection Agency, past and present*)

Environmental Protection Agency, past and present

Prior to 1970, the U.S. Departments of Interior and Agriculture handled our pollution problems at the federal level and the U.S. Congress occasionally dabbled in environmental legislation. By the late 1960s, Americans had come to the conclusion that the health of our planet was intimately tied to human health and welfare, and that federal dabbling simply wouldn't suffice. A few months after the first *Earth Day,* held on April 22, 1970, plans to create the Environmental Protection Agency began.

In January 1971, the EPA was established to "strive to formulate and implement actions that lead to a compatible balance between human activities and the ability of natural systems to support and nurture life." With a budget of $3 billion, a work force of 7,000 employees, and William Ruckelshaus as the first administrator, the EPA opened shop.

The EPA's first decade was one of momentum and progress. Some of its accomplishments were as follows: 1) In 1970, Congress amended the *Clean Air Act* of 1963, giving the EPA the responsibility of setting air quality standards for each pollutant. 2) In 1972, Congress passed the Water Pollution Control Act, which allowed the EPA to set water quality standards and regulate *water pollution.* The EPA aggressively reported offenders to the Department of Justice. 3) In 1976, The *Resource Conservation and Recovery Act (RCRA)* was passed by Congress, which directed the EPA to get involved in preventing industrial *hazardous waste* problems.

In 1978, **Love Canal** introduced the American public to the magnitude of hazardous waste problems that could not be prevented since they already existed. The public began to realize there were thousands of other "Love Canals." In response, Congress passed CERLCA (the Comprehensive Environmental Response, Liability, and Compensation Act—usually referred to as the **Superfund**) in 1980. This fund was meant to help pay for the management and cleanup of these hazardous waste sites. The EPA was no longer simply responsible for preventing pollution, but also cleaning it up. A National Priority List of cleanup sites was created and things began to change at the EPA and with the public's perception of the EPA.

The EPA's progress was halted and often regressed beginning in 1980 under a new administration. Then-President Ronald Reagan felt environmental well-being was contradictory to economic well-being and went to work reversing much of what the EPA had accomplished. The Office of Management and Budget (OMB) was instructed to review the economic impact of all EPA-proposed regulations. The newly appointed EPA administrator, Anne Gorsuch, cut the EPA budget by 50 percent and reduced the staff by over 25 percent. The number of environmental violations reported to the Justice Department dropped by almost 70 percent in the first two years of this administration.

In 1984, however, Congress re-enforced the RCRA and, in 1986, it extended and increased funding for the Superfund and passed the **Clean Water Act.** Responding to public concerns about the well-being of the EPA and the state of the environment, William Ruckelshuas returned in the role of EPA administrator in an effort to regain public confidence and improve internal morale.

Major changes were made during the mid 1980s that concerned many environmentalists. The method in which **risk assessments** of hazardous wastes were determined was changed to reflect an increased willingness to accept certain levels of danger based on the economic worth of the substance.

In 1989, William Reilly was appointed EPA administrator, which pleased most **environmentalists.** His accomplishments were many at first, but shortly after he started, his progress was thwarted by the White House administration. Disagreements with the administration became public during the 1992 **Earth Summit** in Brazil when he and the president had differing opinions on what bills and treaties should and should not be signed.

During the late 1980s, the Bush Administration elevated the importance of economics over the environment by using the

President's Council on Competitiveness as a weapon. This council, along with the Office of Management and Budget, reviewed and rewrote environmental regulations (even though members of the council have no environmental expertise) it felt were unfair to business. The council rewrote pollution standards, allowing industry to increase levels of pollutants whenever they deemed it necessary. The council rewrote the definition of a **wetland** to exclude thousands of acres identified as such by scientists, and stopped a proposal to recycle auto batteries, a major cause of **lead poisoning**.

At the close of 1992, the EPA had 10 regional offices, a staff of about 17,000 employees, and a budget of about $6 billion.

environmental science The **environment,** environmental science, and **ecology** are often used interchangeably. Although similar, they differ. The term environment is usually used from the perspective of an organism. For example, a willow tree's environment refers to everything, living and nonliving, that affects the willow tree. Environmental science pertains to how human intervention affects "our" environment. Ecology is a science concerned with the relationships that exist between all the parts, both living and nonliving, of an environment.

More specifically, environmental science deals with the effect of human populations and technologies on our planet and how to resolve problems they pose. It is an interdisciplinary study that transcends many other sciences, including biology, geology, chemistry, and many others. (See *Environmentalist, Ecology studies*)

environmental tobacco smoke (ETS) The inhalation of cigarette smoke by nonsmokers is called environmental tobacco smoke, or ETS. (It is also called passive smoking.) The Environmental Protection Agency, National Research Council, and the Surgeon General's office have all performed studies that show ETS causes illness and probably death. The most recent study by the EPA, which refers to 50 other studies, draws the following conclusions: ETS causes 3,000 lung cancer deaths per year, contributes to at least 150,000 respiratory infections in babies, triggers at least 8,000 new cases of asthma in unaffected children each year, and aggravates symptoms in at least 400,000 asthmatic children. (See *Indoor pollution, Air pollution*)

environmentalist An environmentalist is someone who understands his or her environment and uses this understanding to help manage our planet. Managing the planet involves lifestyle changes, activism, and voting.

Changing personal lifestyles may seem to have little impact on correcting major problems, but when multiplied by a million people, the impact can be awesome. Individuals who buy *recycled* or biodegradable products not only change their lifestyle, but are changing the way companies do business. *Green products* and new companies are springing up in response to this new market.

In a democratic society, people not only can take action personally, but can also help to create the laws that take action on a larger scale. Legislation such the *Clean Air Act* and the National Environmental Policy Act were enacted due to the pressures applied by environmentalists. You can exert pressure to pass legislation by communicating with your officials, either on an individual basis or by joining *environmental organizations*.

Possibly the best way to help manage the earth is to express your feelings at the ballot box by voting for representatives with views similar to yours or on environmentally sensitive referendums. To do this effectively, environmentalists must be able to detect *greenspeak* and *greenscams*. (See *Environmental literacy, League of Conservation Voters, Environmental education, Eco-conservative*)

estuary Estuaries are unique coastal ecosystems where fresh water from a river mixes with salt water from the sea. This results in a concentration of salts between that of freshwater and marine habitats, called brackish waters. Since the degree of *salinity*, temperature, and other factors varies in estuaries with the tides, only certain kinds of organisms with wide *tolerance ranges* usually inhabit this type of ecosystem.

Estuaries are among the most productive ecosystems on earth, along with *tropical rain forests* and other types of *wetlands*. The constant flow of water from the river or other body of water into the estuary provides high concentrations of *nutrients* for organisms to thrive. The water is usually shallow, so sunlight can reach the bottom, allowing *emergent plants* and algae to grow abundantly. These *producers* are the first link in *food chains* that include fish such as flounder, and crustaceans such as shrimp. Many organisms use estuaries as a spawning ground and a nursery. The young have plenty to eat and protection from the open ocean. Once large enough, they move out to sea, making estuaries a vital link in much larger ecosystems. (See *Wetlands, Estuary and coastal wetlands destruction*)

estuary and coastal wetlands destruction

Estuaries and *coastal wetlands* (bays, lagoons, salt marshes, and swamps) were thought by many, until recently, to be worthless, mosquito-infested regions. Many of these areas were used as dumping grounds for waste, while others have been drained, filled-in, and built upon, and still others have had their source waters diverted for human use.

In fact, estuaries, swamps, and marshes are considered the most productive of all habitats and similar in *net primary production* to *tropical rain forests.* These regions are spawning and nursing grounds to 70 percent of this country's commercial fish and shellfish, and they are the breeding grounds and habitats for many waterfowl and other wildlife.

These areas filter out and dilute water pollutants from the rivers and streams that feed them, before reaching the open sea. Some estimates value one acre of tidal estuary to be worth 75,000 dollars in waste treatment costs. These regions also act as an enormous buffer, protecting the inland from storm waves and absorbing vast quantities of water that would otherwise move on shore, resulting in flooding.

In the U.S., 55 percent of all estuaries and coastal *wetlands* have been damaged or totally destroyed. Most have been filled for building developments due to the popularity of coastal living. Many others are contaminated with toxins due to either direct dumping or the accumulations of toxic substances from the incoming waters.

These regions are still threatened by development. Although few people disagree with scientists about the importance of these habitats, many people still feel these areas should be developed. Some local, state, and federal agencies are attempting to accomplish this by redefining what a wetland is and offering a "no net loss" policy, where manmade environments elsewhere would replace natural wetlands that are destroyed. (See *Everglades, Aquatic ecosystems*)

ethanol
Ethanol is a *biofuel* created by the fermentation of sugar or grain crops, used as an alternative to gasoline. Mixed with gasoline, it is called gasohol. (See *Biomass energy*)

ethics, environmental
Ethics is a branch of philosophy that attempts to determine right from wrong, without cultural influences. Environmental ethics is specifically concerned with defining right from wrong as it pertains to environmental issues. Environmental ethics, like any ethical problem, is defined in differ-

ent ways depending upon the perspective of the individual espousing the beliefs.

Environmentalist **Aldo Leopold** wrote in his essay "The Land Ethic," in 1949, "All ethics so far evolved rest upon a single premise: that the individual is a member of a community of interdependent parts. The land ethic simply enlarges the boundaries of the community to include soils, waters, plants, and animals, or, collectively, the land". (See *Philosophers, environmental*)

ethnobiology   The study of the use of plants and animals by humans.

eugenics   The science of improving the genetic qualities of organisms by selective breeding. Crops such as alfalfa, corn, and cotton have been selective-bred to resist numerous kinds of diseases that would otherwise destroy these crops. (See *Resistant crops, Organic farming, Integrated pest management*)

euphotic zone   Refers to that portion of a body of water that receives sunlight. (See *Aquatic ecosystem, Marine ecosystem, Standing water habitats*)

euroky   The ability of an organism to tolerate a wide range of environmental conditions. (See *Coastal wetlands*)

eutrophication, cultural   Cultural eutrophication is an accelerated form of **natural eutrophication.** The natural process is accelerated by waste products produced by human activities; these waste products enter bodies of water. Untreated **sewage,** livestock wastes, agricultural and domestic **fertilizers,** and many industrial waste products that find their way into bodies of water dramatically hasten the normal process of nutrient enrichment.

This dramatic enrichment of **nutrients** produces a population explosion of organisms such as algae. When these organisms die, bacteria have a population explosion of their own, feeding on the decomposing algae. This results in a reduction in the oxygen content of the water and the death of the lake or pond **ecosystem.**

Symptoms of eutrophic bodies of water (those with high levels of nutrients) include large amounts of shore vegetation, **algal blooms,** stagnant water, and the lack of cold-water fishes. (See *Eutrophication, natural; Biological oxygen demand, Succession, primary aquatic*)

eutrophication, natural Lakes are classified as either oligotrophic or eutrophic. Oligotrophic lakes are deep, clear, cold, and contain limited nutrients to support life. Eutrophic lakes have undergone the process of eutrophication and are usually shallow, warm, and cloudy, since they are rich with **nutrients.**

Bodies of standing water go through natural eutrophication, the process of nutrient enrichment, over long periods of time, usually thousands of years. Sediment is naturally washed into bodies of water from the surrounding **watershed** and the nutrients within this sediment dissolve in the water. High concentrations of nutrients result in vast growths of algae. Over time, enormous quantities of dead and decaying organic matter (dead algae) settle to the bottom, where they provide food for vast numbers of **bacteria.** As the bacteria decompose the dead algae, they deplete much of the dissolved oxygen from the water, resulting in the collapse of the **food web.** Natural eutrophication is part of the process of **succession,** which is the gradual change from one type of **habitat** to another. (See *Cultural eutrophication*)

Everglades The Everglades is a unique **wetlands** ecosystem 200 miles long, by 50 miles or less wide, by a few inches deep, found between Florida's Kissimmee River and the Florida Keys. Coined the "river of grass" by activist and protector **Marjory Stoneman Douglas,** it contains sawgrass, tree islands, and marshes. Found in this region are the 1.5 million-acre Everglades National Park, Big Cypress National Preserve, and the Loxahatchee National Refuge. About 40 percent of these wetlands have been lost to development and the quality of what remains has been degraded due to waste water and agricultural water runoff containing **fertilizers** and **pesticides.** (See *Estuary and coastal wetlands destruction*)

evergreen Refers to trees and shrubs that have leaves year-round, such as pines and spruces. (See *Deciduous, Taiga*)

exobiology The study of life from other worlds. (See *Extinction and extraterrestrial impact*)

exotic species Refers to an organism that is introduced into a new area; one that is not native to that area. Exotic species are sometimes introduced into an area for a purpose, such as during **biological control** of insect pests. In other circumstances, attempts are made to prevent exotic species from entering areas, as with the zebra mussel in waterways, since they soon overrun existing populations and damage stable ecosystems.

experimental ecology   Experimental ecology became popular in the 1960s and continues to be important today. It concentrates on studying the mechanics of an organism's environment by manipulating the environment (or organism) to see what happens when controlled factors are changed. Experimental ecology was preceded by **descriptive ecology** and the predecessor to **theoretical ecology**. (See *Ecology studies*)

extant   Organisms that are living during the present time; not **extinct.**

extinct   Species that are no longer in existence are referred to as extinct. Species are forced into extinction when they cannot adapt to changes in their environment. These changes can occur naturally or can be caused by human activities.

The most likely organisms to become extinct are those that have a low population density, are found in a relatively small area, assume a specialized **niche,** and reproduce slowly. The **Northern Spotted Owl** is a good example of this type of organism. Rabbits, mice, and most insects have a high population density, are found widely, filling many different niches and reproduce rapidly, so they are less likely to be forced to extinction.

Human activities have forced many species into extinction, including the classic example of the passenger pigeon many years ago and the Dusky Seaside Sparrow more recently. Human activities responsible for extinction (and the estimated percent attributed to this activity) are as follows: destruction of an organism's habitat (30%); commercial hunting (21%); human introduction of **exotic species** that competed with the extinct organism (16%); sport hunting (12%); pest control (7%); pollutants (1%); and the remainder for miscellaneous reasons. (See *Biodiversity, loss of; Anthropogenic stress, Endangered species*)

extinction and extraterrestrial impact   The geological records reveal that throughout the earth's history, there have been five mass extinctions. There is a growing body of evidence that an object from outer space collided with Earth on three occasions in the past, disrupting global climate, which could have resulted in three of these mass extinctions. Much of this evidence is based on the find of unusually shaped "microtektites." These are tiny glass beads believed to have formed as a result of the collisions.

The most famous of the three extinctions occurred about 65 million years ago, resulting in the annihilation of the dinosaurs and about two-thirds of all marine life. The other two occurred about 200 and 370 million years ago.

Theories about how the impacts caused mass extinctions center around dust clouds believed to have blocked out the sun enough to disrupt plant growth and destroy *food chains.* The earth might have also cooled sufficiently to disrupt life as it existed at that time. Other theories propose some of the extinctions were due to *global warming* caused by large amounts of carbon dioxide released by limestone that melted during the collision.

Critics of these impact theories believe the mass extinctions were probably caused by natural causes, such as volcanoes that spewed ash, blocking the sunlight and changing the climate. (See *Extinct, Space debris, Ice Ages*)

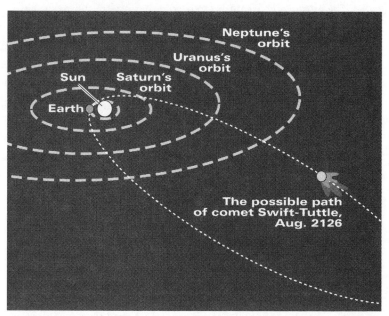

extirpate Refers to the removal of an organism in its entirety; for example, pulling a weed up by its roots.

extremely-low-frequency magnetic radiation (ELF) Power companies in the U.S. and Canada deliver electricity at a frequency of 60 hertz. **Electromagnetic radiation** created by currents at this frequency level is called extremely-low-frequency (ELF) magnetic radiation. Studies done so far on the harmful effect of this radiation are highly debatable and not conclusive. Some American and Swedish studies have indicated that long-term exposure to high-power tension lines increases risks of childhood leukemia and cancers from one in every 10,000 to two or three in every 10,000—doubling or tripling the risks. Other studies find no relation between illness and any form of electromagnetic radiation.

The following items are known to give off ELF magnetic radiation: high-tension power lines, building wiring systems, video display terminals (computer screens), televisions (at close range), electric blankets, heating pads, many household appliances, fluorescent light fixtures, and office equipment that have electric motors or transformers.

Although most research has been performed on high-tension wires, a few studies have found a possible link between the use of household appliances such as hair dryers and illnesses. (See *Indoor pollution, Healthy homes*)

Exxon Valdez On March 24, 1989, an oil tanker, the Exxon *Valdez*, went off course, hit a reef and spilled 11 million gallons of oil into Alaska's Prince William Sound, creating the worst tanker oil spill in U.S. history. (The Amoco *Cadiz* spilled five times as much oil off the coast of France in 1978.) The *Valdez* incident polluted about 1,000 miles of Alaska coastline and coated tens of thousands of animals with oil, many of which died. The degree of harm to surrounding **ecosystems** is still being studied. Optimistic predictions state the area will return to normal within five years of the disaster, but others say it will take decades if at all. Studies done in the summer of 1992 revealed oil still present along shorelines. The oil is still contaminating shellfish, and still being eaten by otters and birds, which show ill effects from the ingested oil.

Exxon spent over a year helping clean up the mess and established a damage-claims program for the residents whose livelihoods were affected. The accident cost Exxon about $2.5 billion for cleanup and damages. The Exxon *Valdez* was not a double-hull tanker, which could have prevented the accident.

The National Transportation and Safety Board stated the captain was guilty of drinking alcohol and leaving the bridge before the accident. The ***Clean Water Act*** of 1972, however, which grants immunity to anyone who reports an oil spill to authorities, resulted in the Alaska Court of Appeals providing the captain with complete immunity of any charges. In 1992 the captain was hired by the State University of New York Maritime College to serve as the watch officer during training exercises. The captain's master's license was restored by the Coast Guard, enabling him to pilot any ship. (See *Oil spills in U.S., Oil Pollution in the Persian Gulf*)

factory farming Most livestock raised for slaughter comes from highly mechanized, high-tech, and high-volume factory farms. Factory farms have high-density populations in controlled environments. Animals are often kept in inhumane conditions. Growth hormones and antibiotics are often used extensively to assure rapid, healthy, and substantial growth in the animals. Some of the substances fed or injected into these animals may pose a health risk to consumers eating the meat. (See *Organic beef*)

fallow field Refers to a field that is left idle to restore its productivity. (See *Soil conservation*)

farmland lost *Urbanization* swallows up about 2.5 million acres of farmland in the U.S. annually. Most of this land is in **high-market value farming counties,** meaning it is some of the most fertile farmland that exists. Some states tax land on its "highest value use." The highest value of this land is often for development. Since many farmers cannot afford to continue farming the land when it is taxed as if it has been developed, the farmers are forced to sell the land for development. Many states have changed the tax laws to only tax on the existing use in an effort to save farmland. (See *Urban sprawl, Urbanization, Soil conservation*)

fast-food packaging After intense pressure from concerned environmental groups and individuals, McDonald's Corporation replaced its *plastic* foam "clamshell" hamburger box with paper packaging in late 1990. (*Paper recycling* programs have many advantages over *plastic recycling.*) Other fast-food chains followed in McDonald's footsteps. More recently, some fast-food chains are attempting to minimize the amount of waste with *source reduction.*

fauna All the animals in a *habitat.*

faunula   All the animals in a microhabitat, such as in leaf litter or under a rock.

Federal Insecticide, Fungicide, and Rodenticide Act (FIFRA)   Created in 1972, this law gave the *EPA* authority to test and regulate *pesticides* in an effort to protect the environment. This law is considered by many to be the most poorly enforced environmental law in effect. Enforcement of this law has long been a battleground between the chemical industry and environmentalists. (See *Pesticides dangers, Pesticide residues, Circle of poison*)

fertilizer   Refers to a substance added to the soil to supply *nutrients* required for plant growth. It can be an *organic fertilizer* or a synthetic fertilizer. Fertilizers supply the three primary plant nutrients: potassium, phosphorus, and nitrogen, called *macronutrients.* They may also contain substances required by plants in smaller amounts, called *micronutrients,* including manganese, boron, zinc, and many others. (See *Monoculture, Green manure, Night soil*)

Flipper Seal of Approval   Earthtrust has created the Flipper Seal of Approval, which shows a cartoon likeness to the celebrity dolphin. Its purpose is to assist consumers in finding tuna that has not been caught at the expense of dolphins in *purse seine nets.* Only those canners using tuna captured following strict guidelines set by the *Earth Island Institute* become certified to use the Flipper Seal of Approval. Earthtrust is based in Hawaii. (See *Seal of approval, Green marketing, Green products, Driftnets*)

flocculation   The process in which clumps of solid waste (flocs) increase in size to expedite their removal. (See *Water treatment*)

floodplain   Refers to lowlands alongside a river, prone to flooding. Some may flood annually, while others only during catastrophic storms, or any duration in between. *Urban sprawl* often uses floodplains for development. (See *Running water habitats*)

flora   Refers to the plant life in a given **habitat.** (See *Niche*)

flowers   Refers to the specialized reproductive organs of flowering plants (angiosperms), usually with female organs in the center, surrounded by male organs. (See *Sexual reproduction*)

fluff   Refers to the nonmetal parts of a car. (See *Automobile recycling*)

fluorescent lighting   Fluorescent lights use less energy to produce more light, than conventional incandescent light bulbs. A 40-**watt** fluorescent bulb produces 80 lumens of light per watt, while a 60-watt incandescent bulb produces less than 15 lumens of light per watt. Over a seven-hour period, the fluorescent bulb in this example saves 140 watts of energy over the incandescent bulb. (See *Radio-wave light bulb, Bioluminscence, Batteries*)

fluvial   Inhabiting streams and rivers. (See *Running water habitats*)

fly ash   Refers to air emissions produced by **incineration.** (See *Municipal solid waste, Air pollution*)

flyways   The **migration** route of birds is called a flyway. In North America there are four major flyways: Atlantic, Mississippi, Central, and Pacific. Migratory birds, such as ducks, geese, swans, and rails, can fly thousands of miles between where they are hatched, often in Canada, and where they winter, usually in Mexico. (See *Tagging programs*)

food chain   The supply of food in any given **ecosystem** can be traced to create a hierarchy of "who eats what." This hierarchy, in its simplest form, creates a food chain with green plants on land or phytoplankton in the oceans (**producers**) at the beginning of the chain. The next link contains animals that eat the plants, called primary **consumers.** They in turn are eaten by other animals called secondary consumers. Tertiary consumers or, possibly, quaternary consumers are usually the last link in most ecosystems. Food chains are the single lines of threads in a complex **food web.** (See *Energy pyramid, Symbiotic relationships, Predator-prey relationships*)

food irradiation   Food irradiation refers to bombarding food products with gamma **radiation** to kill pests and pathogens such as salmonellae and extend shelf life. This method has been used in many countries and has been approved for use in the U.S. Irradiated food does not become radioactive, and most short-term studies indicate the food to be safe for consumption. Many people contend the long-term harmful effects have not been realized and urge irradiated food be banned. Although food irradiation uses radioactive cobalt instead of uranium, as used in **nuclear reactors,** it still results in **radioactive waste** that must be disposed of. Irradiated foods are labelled with the **radura.** (See *Nuclear waste disposal*)

food webs   *Food chains* are connected to other food chains. These intertwined chains create a complex food web that shows all the feeding relationships that exist within an **ecosystem.** For example, a single type of plant might be eaten by five types of insects, each leading to their own food chain, or a mouse might be eaten by a fox or a hawk, each leading to their own food chain.

There are two basic kinds of food webs: grazing and decomposing (also called detritus) webs. The examples mentioned above are grazing webs, in which the energy harnessed by the green plants is used as food to build larger and more complex organisms. In decomposing food webs, however, dead organisms are broken down and decomposed by insects, bacteria, and fungi. For example, a dead mouse might be eaten first by scavenging insects. Once it is broken down substantially, it may be attacked by bacteria and other microbes that complete the decomposition, returning the nutrients to the soil or water. (See *Predator-prey relationships, Symbiotic relationships, Decomposers*)

forest management, integrated   The timber in forests is a **natural resource,** which is exploited for human consumption. The **wildlife** in forests is also a natural resource and should be protected from human destruction. Balancing the economic needs with environmental interests is called integrated forest management. Proper forest-management techniques can minimize the negative environmental impact of logging and sustain many forests indefinitely. Improper techniques can damage or destroy entire forest ecosystems.

Some methods, such as **clear-cutting,** are economically beneficial to the logging industry but are usually devastating to the forest. **Selective harvesting** and **reforestation** are methods that can sustain forests, under some circumstances, while being logged. (See *Forest service, past and present*)

Forest Service, past and present In 1891, the U.S. Forest Reserves were created and managed by the Department of Interior. In 1892, Gifford Pinchot thought that the forests were not managed properly and pursuaded Congress to turn these reserves over to the Bureau of Forestry, which he managed, within the Department of Agriculture. Pinchot then renamed the Bureau of Forestry the Forest Service.

For a good part of the Forest Service's 100-year history, a battle has raged over its primary function: preservation or timber sales. Numerous acts have been passed to reform the Forest Service. Some were in the best interest of preserving the land, while others gave the forest managers complete control over timber sales with little regard to protection. The battle goes on, but the forests are losing and the logging industry is certainly ahead.

The Forest Service today is in the process of finishing fifty-year management plans for each of our national forests, which were ordered by the National Forest Management Act of 1976. All indications reveal that the service plans to favor timber cutting and road-building over protection. This is leading to the destruction of the nation's most valuable resources. Roadbuilding and **clear-cutting** destroy habitats and wildlife and cause **soil erosion** and stream pollution.

During the past five years, more than half of our national forests have actually lost money selling timber. However, Forest Service managers still receive budgetary rewards for selling timber even if they lose money. The Forest Service spent over $600 million more to manage its timber sales program than it made from the sale of trees. The American taxpayers are financing the destruction of their own national forests.

According to environmentalists, most national forest land is too remote and does not produce enough high-quality timber to justify the cost of cutting down these trees. Not only are we losing money but destroying the habitat as well. Many environmental organizations are trying to make Congress and the public aware of these issues and establish a proposal for Forest Service reform. Some of these goals are to adopt a moratorium on Forest Service road construction, establish a process for eliminating below-cost timber sales, abolish the timber pricing system that allows the timber buyer to profit but ignores the taxpayer's cost, and to charge fees for hunting and fishing to give managers incentives to protect the wildlife and preserve the forest's natural beauty. (See *Association of Forest Service Employees for Environmental Ethics, Forest management, integrated*)

formaldehyde   Formaldehyde is a **volatile organic compound (VOC).** It is used in many building materials, such as particleboard, plywood paneling, and fiberboard. It is also used extensively in **plastics,** textiles, **pesticides,** and cosmetics. Its popularity is primarily due to its ability to bind dissimilar substances together to produce new, complex compounds. Over 6 billion pounds of this colorless gas were produced in the U.S. in 1989.

Formaldehyde "evaporates" out of the products mentioned above by a process called **outgassing.** Most outgassing occurs within the first few months of production, releasing substantial quantities of the substance into the air. Formaldehyde is often found within buildings in fairly high concentrations.

The most obvious health problems associated with formaldehyde are upper respiratory tract irritations producing symptoms similar to a cold, irritated eyes and sinuses, coughing, and running nose. Some people get headaches. It has been identified as a "probable human carcinogen," but this issue is still being researched and debated. (See *Indoor pollution, Healthy homes, Carcinogen classification*)

formations   Scientists have divided the earth into large regions containing distinct groups of plant life called formations. For example, all the plants in a grassland are considered a formation. (See *Biomes, Zones of life*)

fossil fuel   Fossil fuels include coal, oil, and natural gas. These fuels provide the majority of the world's energy supply. They are considered **nonrenewable energy** sources, since deposits of these substances are no longer being replenished and will become depleted in the future. (See *Oil, Coal, Natural gas, Oil formation, Coal formation, Alternative energy*)

fossil fuel reserves   Reserves are the identified quantities of fossil fuels available for immediate extraction from the earth. Reserves are that portion of the total amount believed to exist within the earth that are economically feasible to extract. For example, China has over 50 percent of the world's total remaining ***coal*** reserves and what was the Soviet Union has about 40 percent of the ***natural gas*** reserves. (See *Fossil fuel resources, Oil formation, Energy sources, historical*)

fossil fuel resources   Fossil fuel resources refers to the projected amount of a fossil fuel that exists within the earth. The total resource may not ever become available to use, since it might not be economically feasible to extract, or the technology to do so might not exist. (See *Fossil fuel reserves, Coal formation, Energy sources, historical*)

freshwater ecosystems   ***Aquatic environments*** can be divided into ***marine ecosystems*** (salt water) and freshwater ecosystems. Fresh water contains relatively low concentrations of dissolved salts and therefore provides a very different habitat from marine bodies of water. Freshwater ecosystems can be divided into two types: ***standing water habitats,*** which include ponds, lakes, and reservoirs; and ***running water habitats,*** which includes streams and rivers. Limnology is the study of all types of freshwater habitats.

friability   Refers to a soil's ability to crumble. Friability is necessary for good plant growth. Friability is usually determined by the ***soil texture*** and the amount of moisture in the soil. For example, "sandy" soils are friable but "clay" soils are not. (See *Earthworms*)

Friends of the Earth   This organization works with grassroots environmental groups, lobbies members of Congress, conducts workshops, and provides advice and information to the public. It focuses on major environmental problems such as ***ozone depletion*** and ***toxic waste,*** but also concentrates on energy conservation, ***green taxes,*** and corporate accountability. In 1990, it joined forces with two other groups: the Environmental Policy Institute and the Oceanic Society. Write to Friends of the Earth, 218 D Street SE, Washington DC 20003.

fruit waxing Cucumbers often have a layer of wax applied to their surface to prevent shriveling and improve appearance. This wax is often embedded with **fungicides** to prevent mold growth. The wax also seals in any other **pesticides** that were applied to the plant before waxing. Many other types of produce are also routinely waxed but not as noticeably as cucumbers since a thinner coat is used. Apples, bell peppers, citrus fruits, eggplants, squashes, and tomatoes are all candidates for waxing.

Waxed produce is required by federal law to be marked as such. Although these signs rarely appear, stores can be fined 1,000 dollars for failure to do so. Consumers can request these signs be posted in their supermarket. If this doesn't work, report the violation of this Food and Drug Administration (FDA) law to your state attorney general's office. For more information about safe foods, contact Americans for Safe Food, 1875 Connecticut Ave NW, #300, Washington DC 20009. (See *Pesticide dangers, Pesticide residues on food*)

fungicide Refers to a **pesticide** that kills **fungi.** Fungi help decompose dead organisms or tissues naturally. Most fungicides are applied as fumigants to protect stored food products from spoiling.

fungus Fungi are primitive plants, incapable of photosynthesis. Most reproduce **asexually** by spores. Since they cannot produce their own food, they are either **parasitic** on other organisms (such as the fungus that causes athlete's foot) or saprophytic, meaning they feed on decaying plants (such as mushrooms).

Gaia Hypothesis   Gaia was the ancient Greek Mother earth deity. The Gaia Hypothesis, proposed in 1972 by Lovelock and Margulis, suggests that all life on earth acts together in a way that causes the planet to be self-regulating. The hypothesis states that the planet possesses a life force that monitors and maintains conditions at a level best suited for the continued survival of life on earth. Some people believe this means that any single species threatening the survival of life in general on the planet would naturally be forced into extinction. (See *Deep ecology, Ethics, environmental; Extinct*)

gall   Refers to an abnormal growth, usually a swelling on the leaves or twigs of a plant, caused by the invasion of either microbes, **fungi,** or **insects.** The invaders use the structure for protection and as a food source. The plant is usually unharmed. (See *Symbiotic relationship*)

garbage   Garbage is any substance that is no longer needed and must be disposed of. It may be anything from leftover food to an old refrigerator or automobile. Garbage is similar in many ways to a **pest** or a **weed** since all three can be considered useful if found in another place at another time. One person's garbage might be another person's food, just as one person's weed might be another person's flower. Since most garbage now finds its way to a municipal collection site, it is commonly referred to as **municipal solid waste** (**MSW,** for short). (See *Garbage magazine, Recycling*)

Garbage magazine   *Garbage: The Practical Journal for the Environment,* is a bimonthly publication available on newsstands and by subscription. One look at an issue can make anyone aware of how garbage, and what we do with it, is of paramount importance to our planet. Each issue contains many one- and two-page articles, so you're bound to find one or more of interest to you. You can reach the magazine at (800) 274-9909. (See *Municipal solid waste, Sewage, Recycling, Garbage*)

gasification   One method of using **biomass energy** involves converting biomass (plants and animal wastes) into liquid or gas fuels called *biofuels.* The process is called gasification, if the fuel produced is in the form of a gas. One example of gasification combines heat, steam, and small amounts of oxygen along with the biomass to produce a mixture of carbon monoxide and hydrogen called *syngas.* This is short for synthetic natural gas, since it has many of the same properties as natural gas.

gasohol   Refers to a mixture of 10 to 23 percent gasoline mixed with the *biofuel* ethanol.

GATT (General Agreement on Tariffs and Trade)   GATT has been in existence since 1974, but in recent years has become a threat to many environmental laws passed in the U.S. GATT was designed to promote free trade among nations. One way it promotes this trade is by urging countries to standardize their laws and regulations with other countries so no trade barriers exist. In many instances, this means the U.S., with relatively strict environmental laws compared with some other nations, must relinquish their standards to comply with international standards, which are usually lower.

The U.S. Marine Mammal Protection Act, which was passed by Congress, prohibits the sale of tuna caught in *purse seine nets,* since thousands of dolphins are indiscriminately killed in these nets. Mexico sued the U.S. under the GATT agreement, as a barrier to free trade. Congress must approve the complaint before the law can be overwritten. Another environmental issue under attack by GATT is the amount of *pesticide residues* allowed on foods sold in the U.S.

geology   The study of the earth, especially regarding the history of rocks. (See *Oil formation, Coal formation, Panaspermia*)

geothermal energy   Geothermal energy, along with solar, wind, hydro, and biomass sources, is called a *renewable energy* source since the supply is considered inexhaustible. The inner core of the earth consists of a molten mass that acts as the source of geothermal energy. In some areas in the U.S. and throughout the world, the intense heat within the earth occurs near the earth's surface and heats underground water, forming hot water or steam.

If these reservoirs are close enough to the surface, wells can be drilled to tap the steam and hot water. The steam and hot water is used to produce electricity with generators. (Geysers occur where these reservoirs of steam and hot water naturally break through the surface.)

About 15 countries use geothermal power, generating electricity at about the same cost as using coal. The largest geothermal plant is in northern California, which supplies energy for the city of San Francisco. Reykjavik, the capital of Iceland, uses geothermal power to heat all commercial buildings.

Geothermal power is relatively clean, but it does create some environmental problems. It can cause **air pollution** since it emits hydrogen sulfide and ammonia, and can even emit some radioactive substances. It also can cause **water pollution** from some dissolved solids it brings to the surface. Pollution control devices are used to prevent environmental damage.

Gibbs, Lois   In 1978, Lois Gibbs discovered a **toxic waste** dump site at an abandoned chemical plant three blocks from her home in Niagara Falls, New York, in a neighborhood known as the **Love Canal.** She believed this dump site might be the cause of the severe asthma, seizure disorders, and other illnesses in her children and others in the area. Although she passed out petitions, gave speeches, and lobbied politicians to clean up the site, her plea was ignored. Finally she took a few **EPA** agents hostage, which got the attention of then-President Jimmy Carter. This led to the government's evacuation of the neighborhood. Currently, Ms. Gibbs is a Washington lobbyist, involved in the Citizens' Clearinghouse for Hazardous Wastes, which assists people living in communities similar to Love Canal.

gillnets   Gillnets are fishing devices used in coastal waters to catch large numbers of tuna and salmon. Unsuspecting fish swim into these walls of netting, entangling their gills. In addition to the intended catch, the nets entangle large quantities of fish and other marine life not being sought. This unintentional catch, called **by-catch,** is thrown back into the sea, with most of the animals dead or dying. These nets are similar to **driftnets** but are not as large, since they are used in coastal waters.

It's estimated that 2,500 ships put out over 50,000 miles of driftnets and gillnets every night. These nets devastate populations of many fish species, killing dolphins, whales, seals, and sea turtles, among other sea life. (See *Purse seine nets, Turtle excluder devices, Flipper Seal of Approval, Seals of approval, environmental*)

glade   A natural clearing in a woodland.

gley   A type of **soil** characterized by periodic **waterlogging** due to poor permeability.

Global   Environment   Monitoring   System (GEMS)   GEMS is a worldwide effort that monitors the global environment and makes regular assessments about the health of our planet. One hundred and forty-two countries participate in the global data collection and assessment. It was founded in 1972 as the worldwide monitoring and assessment arm of the **United Nations Environment Programme (UNED)**.

global warming   **Greenhouse gases** produce the **greenhouse effect,** which traps heat near the earth's surface, maintaining a relative constant temperature. Many human activities increase the amount of greenhouse gases in the atmosphere, which can result in a gradual increase in the earth's surface temperature, a process called global warming.

   Carbon dioxide is the primary greenhouse gas. It occurs naturally and is vital to life, but excessive quantities of it are released by burning fossil fuels (coal, oil, natural gas). Other greenhouse gases are almost exclusively produced by human activity such as **CFCs,** which are used as refrigerants. Still other greenhouse gases include **methane, nitrogen compounds,** and **ozone.** About 80 percent of global warming is due to increases in all these gases.

   Deforestation is believed to account for the other 20 percent. Plants incorporate carbon dioxide into their bodies during photosynthesis. Fewer trees caused by deforestation mean less intake of carbon dioxide. In addition, burning this wood (along with fossil fuels) sends the carbon dioxide back into the atmosphere at an accelerated rate.

   Since 1880, the worldwide average temperature has increased about .9° F. This, however, is within the normal fluctuation range, meaning it could be a short-term change that will return to normal in the near future. Computer model predictions, however, estimate an increase of between 1.3° F and 20° F in the future. Even the low-end increase could cause dramatic changes in the earth's climate. (See *Slash-and-burn cultivation*)

God Squad, The   See *Endangered Species Act.*

Goodall, Jane   Dr. Jane Goodall has studied chimpanzees for over thirty years. She established the ChimpanZoo program to evaluate and improve the lives of chimpanzees in *zoos.* She also works to eliminate the use of chimps as specimens in scientific research. Currently her time is spent at the Gombe Research Center as well as doing speaking tours to raise money for continued research on chimpanzees.

Grand Canyon   The Grand Canyon, located in northwestern Arizona, is considered one of the seven natural wonders of the world. The exposed rock was eroded by the Colorado river during a process believed to have taken about 10 million years. It was originally protected in 1908 and became a national park in 1919. In 1975, it was expanded to encompass over 1.2 million acres of land. Four million people visit the park each year. (See *National Park and Wilderness Preservation System*)

grant-making foundations   Over $7 billion of grants were donated through philanthropic foundations in the U.S. in 1990. About 5 percent of this total went to environmental causes. The five foundations that contributed the largest amounts to environmental causes that year and the amounts donated (in millions) were as follows: The Richard King Mellon Foundation ($23.5), The J. D. & C. T. MacArthur Foundation ($23.3), The Pew Charitable Trusts ($14.6), The Ford Foundation ($12.9), and the Rockefeller Foundation ($10.9). (See *Environmental organizations*)

grass cycling   One of the most obvious forms of *recycling* is overlooked by millions of individuals who mow their lawns and remove the clippings, only to apply *fertilizer* to replace the lost *nutrients.*

Dead and decaying organisms are the primary source of nutrients for future generations of plants. Lawn clippings have a fertilizer value of 5-1-3 (nitrogen, phosphorus, and potash). It takes about two pounds of fertilizer per thousand square feet to replace the nutrients removed by the clippings. When left on the lawn, clippings decompose and become useable to the growing grass in about one week. This reduces the amount of fertilizer required by about 25 percent. Grass cycling keeps clippings (which are a very bulky "waste") out of *landfills* and lowers the need for problematic fertilizers.

To recycle grass clippings, you should cut the grass often since the clippings must be one inch or shorter and the grass should be left about two inches tall. (Remember, you don't have to bag, so it takes less time.) The grass should not be damp. Short clippings and tall, dry grass assure that the clippings fall between the remaining grass instead of smothering it. (See *Organic fertilizer, Green manure, Lawnmower pollution*)

grassland, temperate   A temperate grassland is one of several kinds of *biomes.* Temperate grasslands are also known as steppes or prairies. They receive only 25 to 75 centimeters (10 to 30 inches) of precipitation per year. They are usually windy environments with hot summers and mild-to-cold winters. Grasses are the *dominant* plants, making up about 75 percent of the total plant life, since they require less water than trees.

Grazing animals (*herbivores*)—from small mice and ground squirrels to bison, wild horses, sheep, cattle, and wildebeest—are often abundant. There are also many species of insects, reptiles, and birds. Grassland *soils* are very rich and fertile and are therefore being converted at an alarming rate into farmland. Some of the drier grasslands have been converted into grazing land for domesticated animals. (See *Monoculture, Desertification*)

gravel   A particle of *sediment* that falls between 2 and 256 millimeters (less than .1 inch to 10 inches) in diameter. Gravel is subdivided into cobble, pebble, and granule. (See *Soil particles*)

gray air smog   Synonymous with *Industrial Smog.* (See *Air pollution*)

grazing   Refers to foraging for food by *herbivores.* The food eaten is either plants such as grasses, in terrestrial *habitats,* or algae and other phytoplankton in *aquatic ecosystems.*

green cities, U.S.   The 1993 *Information Please Environmental Almanac,* which is compiled by World Resources Institute, rates cities according to their environmental quality and awareness. They used 28 indicators to compare 75 metropolitan regions (with populations over 500,000). Categories analyzed included waste, water use and source, energy use and cost, air quality, transportation measures, toxic chemical accident risk, environmental amenities, and environmental stress.

The top fifteen cities are as follows: 1) Honolulu, HI; 2) San Diego, CA; 3) San Francisco/Oakland, CA; 4) El Paso, TX; 5) Washington DC; 6) Austin, TX; 7) Fresno, CA; 8) New Bedford, MA; 9) Tucson, AZ; 10) New Haven, CT; 11) Rochester, NY; 12) San Antonio, TX; 13) Bakersfield, CA; 14) Pittsburgh, PA; and 15) Miami, FL. (See *Green countries, Bicycling, best cities for*)

green countries Many organizations have ranked countries according to their environmental policies, each with their own selection of criteria. *Newsweek* magazine rated 30 countries in 1992 on three of the most important issues: **population, deforestation,** and **pollution.** Three countries got good marks across the board. They were Costa Rica (noted for pioneering protection of **biodiversity** and forests), Israel (noted for pioneering **solar power** and desert agriculture), and France (noted for its low birthrates and low **greenhouse gas** emissions). The U.S. didn't do very well because of high greenhouse emissions (17.8 percent of world total, the highest in the world) and federal subsidizing of deforestation through the U.S. **Forest Service.** (See *Green cities*)

Green Cross Certification Company See *Seals of approval, environmental.*

green manure Green manure is one of three types of **organic fertilizers.** Green manure refers to any plant that is plowed into the soil to improve fertility. It increases the amount of organic matter available to the next crop grown in the field. Typical types of green manure include any weeds that may have grown on an uncultivated field and turned over into the soil, rangeland grasses turned over into soil to be cultivated, or nitrogen-fixing plants such as alfalfa, planted to enrich the soil. (See *Nitrogen fixation*)

green marketing Many manufacturers are taking advantage of a surge in consumer interest of environmental problems. Marketing studies show that consumers prefer to buy products that are environmentally safe and are willing to pay more for them. Advertising that targets these issues is called green marketing or green advertising.

Much of this marketing has little, if any, basis in fact. Some companies have been forced to stop making environmental claims when they were accused of unfair and deceptive practices. For example, ads about **disposable diapers** and plastic garbage bags that were claimed to readily decompose were removed from distribution since their claims could not be substantiated. Both consumers and manufacturers are trying to develop standards by which these claims can be justified.

In late 1990, a council of 11 state attorney general offices issued a number of guidelines called "The Green Report." The guidelines in this report were used voluntarily by some manufacturers to standardize advertising. The Federal Trade Commission has recently issued its own nationwide "voluntary" guidelines. Many of these guidelines are weak. For example, a product labelled "recycled" need have only 1 percent of recycled materials.

A few states, such as New York and California, have passed their own more stringent guidelines. For example, California law requires a product to contain 10 percent post-consumer recycled paper before it can use the term recycled. Some new environmental **seals of approval** are helping consumers make their purchasing decisions. Until standards are created and enforced, caveat emptor (let the buyer beware). (See *Green products, Recycling, Paper recycling, Recycling logos*)

green products   In 1991, 12 percent of all new products manufactured claimed to be "good for the environment." Probably only a fraction of these products had a right to truly be considered "green." Products that are truly environmentally friendly, however, do exist. There are two kinds of green products. "Deep-green" products are primarily sold through companies that specialize in these products. Most of these companies go out of their way to substantiate their **green marketing** claims. These products were developed "from scratch" to be environmentally friendly. Many companies sell deep-green products and most advertise in popular **eco-magazines** such as **E Magazine** and **Buzzworm.**

The other type of green product is "greened-up", meaning it was once sold as a regular product and then converted into a "green" product. Greened-up products are often sold by the big companies. When it comes to green products, the end justifies the means and the only important thing when selecting green products is to demand the claims be substantiated. (See *Seals of approval, environmental; Green marketing, Recycling logos, Radura*)

Green Report, The See *Green marketing.*

Green Seal See *Seals of approval, environmental.*

green tax Refers to taxes imposed on products or activities that pollute, deplete, or degrade the environment. These taxes act as an incentive to reduce the abundance of these products or activities, urging alternatives to be found, and also help pay for the cleanup of harm already done. One such proposed tax is based on the amount of carbon in **fossil fuels** used, called a carbon tax. Other proposed taxes include those on **hazardous wastes** produced, **pesticides** sold, virgin paper used, and water depleted from **aquifers.**

One of the few green taxes already in place is a tax on **CFCs,** which are used as refrigerants and are a major cause of **ozone depletion.** The tax was designed to force industry to look for CFC substitutes and is also expected to generate about $5 billion in federal revenue over five years. Green taxes have been used successfully for many years in Europe. (See *Disposal fees, Bottle bill*)

greenhouse effect The earth's surface temperature is controlled by many factors, including the greenhouse effect, which works something like a glass greenhouse. When radiant energy from the sun reaches the earth's atmosphere, it passes through the **greenhouse gases,** heating the earth's surface. The heat (infrared radiation) is then reradiated (released) from the earth back up into the atmosphere. The greenhouse gases, however, absorb infrared radiation, trapping it and heating up the lower portion of the atmosphere.

As long as the amount of greenhouse gases remains constant, along with other climatic factors, the temperature on the planet remains relatively steady. Increased amounts of greenhouse gases due to human activities increase the greenhouse effect and are believed to lead to **global warming.**

greenhouse gases **Global warming** is believed to be caused by increased concentrations of greenhouse gases emitted by human activities into the atmosphere. Greenhouse gases include **carbon dioxide** from burning fossil fuels, **chlorofluorocarbons (CFCs)** from air conditioners and refrigerators, **methane gas** from landfills and feedlots, and the **nitrogen compound,** nitrous oxide, from burning fossil fuels and fertilizers. Ground-level *ozone,* produced by burning fossil fuels, is also considered a greenhouse gas. (Don't confuse ground-level ozone with the ozone layer that is up in the stratosphere.)

Carbon dioxide is the principal greenhouse gas. Studies show the amount of carbon dioxide in the atmosphere has been rising over the past few decades. Studies in Hawaii show an increase from 315 ppm in 1958 to 350 in 1990. Burning gasoline for automobiles and burning coal and oil to generate electricity are believed responsible for this increase.

Greenhouse gases are produced primarily in *more developed countries.* The U.S. produces about 18 percent of the total emissions each year, followed by European countries with 13 percent. Other major contributors are the former Soviet Union nations, Brazil, China, India, Japan, Canada, and Mexico.

Greenpeace   Greenpeace is a worldwide organization with over 4 million members. It aggressively and sometimes dramatically promotes public awareness about environmental problems and possible solutions. Greenpeace is primarily involved with defending the *marine ecosystems* and controlling the spread of toxic substances and nuclear weapons. They have been relentless in their pursuit to stop the use of *driftnets* and the indiscriminate death of sea mammals and fish.

Their ship *The Rainbow Warrior* has often been used to protest activities such as whaling, driftnet operations, and nuclear testing. *The Rainbow Warrior* was destroyed by the French government during one of these protests. Greenpeace won a lawsuit against the French government for the attack and was compensated for the loss of the ship. Greenpeace continues to be one of the premier worldwide environmental organizations. The annual membership fee is 15 dollars. Write to 1436 U Street NW, Washington DC 20009, or call (202) 462-1177. (See *Environmental organizations*)

greens, the   In general terms, "the greens" refers to any political party that makes the environment its principal concern. There are, however, over 50 parties throughout the world actually called The Greens. The party originated in West Germany and stressed environmental issues and concerns. The American Greens organized during the mid 1980s, and uses The Greens Clearinghouse in Kansas City, Missouri, to disseminate information about local green parties. Call (816) 931-9366. (See *Environmentalist, Eco-conservative, League of Conservation Voters*)

greenscam As defined by the **League of Conservation Voters,** a "crime committed by politicians who portray themselves as environmentalists, but consistently vote to pollute our air and water, and destroy our planet." (See *Green products, Green marketing*)

greenspeak Refers to politicians' rhetoric about being pro-environment when their voting record is clearly anti-environment. (See *Greenscam*)

greenways Refers to linear parks usually containing a trail for **bicycles** or hiking. These narrow parks are often built along parkways, rivers, or on converted railroad tracks. Besides providing recreation, some are used as commutation routes. (See *Rails-to-Trails Conservancy, Urban open space*)

groundwater Water found on the surface of the earth is called **surface waters.** Water that is absorbed into the earth's crust and stored in underground **aquifers** is called groundwater. Only .5 percent of all water is groundwater, but it supplies much of the world with fresh drinking water. About 50 percent of all drinking water in the U.S. comes from groundwater. Groundwater becomes depleted in a process called **water mining. Groundwater pollution** is becoming a major environmental problem.

groundwater pollution **Groundwater** is a vital resource that is being both depleted and polluted. It is polluted in four ways. 1) Agricultural products, especially **pesticides,** have already contaminated groundwater in 34 states. Seventy-three pesticide residues have been found in groundwater. Fertilizers and animal feedlot wastes also change the chemistry of some groundwaters.
    2) Landfills leak **leachate,** which carries our landfill waste products to our water supply. Ninety percent of all landfills do not catch leachate from passing to the groundwater supply. 3) **Hazardous waste disposal sites** leak contaminants into the groundwater. There are 1.4 million "known" underground storage tanks holding hazardous substances and over 25 percent of them are known to be leaking their contents, with much of it reaching groundwaters. There are also about 180,000 pits and lagoons used in the U.S. to store wastes. Almost all are unlined and most have never been inspected for leaks. Many are believed to be leaking and contaminating groundwater. 4) Well-designed, properly functioning

septic tanks should not cause harm, but millions of the 20 million in use don't function properly, thereby contaminating groundwater.

Group of 10, The   The Group of 10 is an informal name used within the environmental community that refers to 10 influential **environmental organizations** that often exchange ideas and information. Included are: **Environmental Defense Fund, The Wilderness Society, Sierra Club,** Sierra Club Legal Defense Fund, **National Audubon Society,** National Parks and Conservation Association, National Resources Defense Council, **Friends of the Earth, National Wildlife Federation,** and the **Izaak Walton League of America.**

growthmania   A phrase coined by economist E. J. Mishan to describe the belief that "bigger is always better." (See *Economy vs. environment, Economic vs. sustainable development*)

guano   Refers to the accumulation of large deposits of bird, bat, or other animal excrement. Guano is rich in organic **nutrients** and can impact the **ecology** of the **habitat.**

guild   All the species within an ecosystem that utilize the same resource in the same way are described as a guild. For example, many different types of insects feeding on the leaves of trees form a guild and therefore can be studied as a single unit. (See *Competition*)

gully reclamation   **Soil erosion** occurs quickly on sloping land without sufficient plant cover. Runoff water forms gullies that wash away the soil. Gully reclamation returns this eroded land back to productivity. Gullies are planted with fast-growing crops such as oats and wheat. If the water movement is too great, mini-dams are created and allowed to fill with sediment, which are then planted. Fast-growing shrubs, vines, and trees can also be used to stabilize the soil. (See *Soil conservation*)

gyre   The spiral movement of ocean currents. For example, the South Atlantic gyre is the major anticlockwise circulation of surface waters in the South Atlantic Ocean. (See *Upwelling*)

habitat Habitat refers to the place where an organism lives. The habitat must fulfill the needs of the species if it is to survive. These needs include sufficient nourishment and water, proper temperature, sufficient sunlight, etc.

There are two major categories of habitats: aquatic and terrestrial. Aquatic is divided into freshwater and marine habitats. Freshwater is then divided into *standing water habitats,* which include lakes, ponds, and bogs, and *running water habitats,* which include rivers, streams, and springs.

Marine environments are divided into the *neritic zone,* which is the shallow waters off the coasts, and the *oceanic zone,* which is the deep waters. Terrestrial habitats are divided in *biomes.* When organisms live in restricted areas—such as the confines of a leaf, under a rock, or in a pile of dung—the term microhabitat is often used.

habitat management See *Wildlife management.*

halons This substance is used in fire extinguishers and is an *ozone-depleting gas.* Although produced and released into the atmosphere in much smaller quantities than *CFCs,* halons are far more damaging to the ozone layer. "Halon-free" fire extinguishers say either "dry chemical" or "sodium bicarbonate" on the label.

Hanford Nuclear Reservation The U.S. military has produced vast quantities of nuclear waste as a by-product of manufacturing bombs. About two-thirds of it (65 million gallons) is stored at the Hanford Nuclear Reservation near Richland, Washington. Most of it is stored in large tanks and drums, but large amounts were simply dumped into pools and lagoons as late as 1970. Dangerous levels of *radiation* have been found along the Columbia River all the way to the Pacific Ocean, 200 miles away. Recent studies have revealed that radioactive iodine, used to reprocess spent *nuclear reactor* fuel rods, was released from the plant during the late 1940s.

The nuclear waste that hasn't made its way into the soil and groundwater is waiting for a permanent *nuclear waste disposal* site, which will probably be deep in the earth's crust. The

Department of Energy has begun a cleanup effort, conservatively estimated to cost the taxpayer about 30 billion dollars. Hanford is one of 15 locations referred to by the Department of Energy as part of the Nuclear Weapons Complex. All of these sites contain nuclear waste and require cleanup. (See *Vitrification*)

hard pesticides Refers to **pesticides** that do not decompose into harmless chemicals readily. Hard pesticides can remain in their original state for 2 to 15 years. (See *Chlorinated hydrocarbons, Insecticides, Pesticide dangers, Bioaccumulation*)

hardiness Refers to the ability of an organism to withstand freezing temperatures.

hardpan Refers to a **soil** condition in which large amounts of clay accumulate and create a layer that water cannot pass through. Hardpan usually occurs in forest soils. (See *Soil texture, Waterlogging*)

hazardous waste Hazardous waste refers to all substances that pose an immediate or long-term danger to the health or well-being of humans or to the environment during transport or storage of these substances. These dangers are defined by the **EPA** in four categories: 1) "Ignitability" refers to waste that can easily catch fire. 2) "Corresiveness" refers to waste that requires special containment since it corrodes normal materials. 3) "Reactivity" means the waste could easily explode. 4) "Toxicity" means the waste can cause physiological harm to humans or other organisms. This final category is often separated from the others and called **toxic waste.**
  Hazardous wastes come from many sources. The hazard might be a product itself, or a by-product of the manufacturing process. Some common hazardous wastes are: organic chlorine compounds from **plastics** and **pesticides; heavy metals** and various solvents from medicines, paints, metals, leather, and textiles; and salts, acids, and other corrosive substances from **oil** and gasoline.
  The improper management of hazardous wastes is becoming an environmental dilemma. There are about 250,000 **hazardous waste disposal** sites located in the U.S., all posing a threat to our **aquifers,** our health and the environment in general.

hazardous waste disposal For decades **hazardous waste** was disposed of by simply being dumped. Sixty-six percent of this waste is dumped into or onto the soil in 75,000 industrial *landfills* and 180,000 ponds or lagoons. In addition, there are countless storage facilities with thousands of corroding steel drums, containing hazardous wastes. Finally, deep-well injection sites deposit the waste 20 feet to several thousand feet into the earth.

All of these soil-based sites are likely candidates to contaminate *aquifers* that supply much of our drinking water. Estimates have about 2 percent of our aquifers already contaminated with hazardous wastes and the number is rising rapidly.

Those hazardous wastes that are not deposited on or in the soil go directly into the surface waters. Twenty-two percent is simply discharged into rivers, streams, or even sewage systems where it still makes its way to the open waters. This waste damages, if not destroys, many *aquatic ecosystems.* Other methods, such as burning in boilers, chemical treatments, ion exchange, and incineration, make up the remaining 12 percent. New methods of hazardous waste disposal are being tested and used in pilot programs.

Hazardous wastes, especially *toxic wastes,* are gradually infiltrating our entire environment and our bodies, and the price to clean them up will be staggering. The federal government has set aside a *Superfund* for cleaning up many of these sites. (See *Hazardous waste disposal, new technologies for; Breast milk and toxins, Pollution*)

hazardous waste disposal, new technologies for *Hazardous waste disposal* usually means dumping the material into a hole in the ground or into a body of water, resulting in soil and water pollution and a health hazard. Alternatives are now being tested and include: waste destruction, waste immobilization, and waste separation. Waste destruction is used on organic hazardous waste materials and can destroy 99.9 percent of the waste by breaking it down to harmless substances. Waste immobilization places hazardous substances into a form that can be easily disposed of and is less likely to leak into the environment, for example, solidifying a block of material that can then be buried without fear of it contaminating the groundwater. Finally, waste separation is used to separate the hazardous materials from nonhazardous materials, thereby reducing the volume of the hazardous waste.

Many environmentalists feel the best solution for hazardous wastes is to use less of them in our society. (See *Remediation of hazardous waste, Bioremediation, Source reduction*)

hazardous waste, military   The U.S. military generates more ***hazardous waste*** each year than the five biggest chemical companies combined. This includes 500,000 tons of toxic substances and millions of tons of contaminated waste water. Ninety-seven military properties have been put on the ***Superfund***'s National Priorities List of most contaminated sites and another 2,000 military installations do not comply with federal environmental laws. Many grassroots organizations and the Military Toxics network in Seattle, Washington, is taking an active role in trying to get the government to clean up these sites and prevent more from occurring. (See *Hanford Nuclear Reservation*)

HCFC   HCFCs (hydrochlorofluorocarbons) are modified versions of ***CFCs.*** HCFCs are replacing CFCs in many products, since it is less harmful to the ***ozone*** layer. Since it is still an ***ozone-depleting gas,*** however, attempts are being made to phase it out of use and find substances that cause no harm to the ozone layer whatsoever. HCFCs are commonly used in ***polystyrene*** foam products.

HDPE   HDPE is an acronym for high-density polyethylene, which is one of the most common types of ***plastic.*** About 30 percent of all plastics produced in the U.S. are made from HDPE. It is used to make rigid containers for items such as milk and motor oil. HDPE is one of the more commonly recycled types of plastics. For ***plastic recycling*** purposes, these products are stamped with the number "2," surrounded by a triangle formed by arrows.

HEAL   The Human Ecology Action League, Inc. (HEAL) was incorporated in 1977 by physicians and citizens concerned about the harmful effects of chemicals in the environment and its threat to human health. HEAL publishes a quarterly newsletter, "The Human Ecologist," which contains information on environmental health hazards, legislative matters, and sources for safe food, clothing, and household supplies. Write to Human Ecology Action League, Inc., P.O. Box 49126, Atlanta, GA 30359.

healthy homes   Homes built following a set of principles proposed by the philosophy of **baubiologie.** These principles include building homes with materials that don't emit toxic substances, using natural building materials, eliminating use of materials that emit **radiation,** and minimizing electromagnetic fields. These building practices have been popular in Europe for many years, but are just beginning to gain popularity in the U.S. (See *Electromagnetic radiation, Indoor pollution*)

heating, ventilation, and air–conditioning systems (HVAC)   Beginning in the 1970s, many buildings were constructed with nonopening windows to save on energy costs. The indoor environment of these buildings is controlled by HVAC (heating, ventilation, and air-conditioning) systems. Fresh air is introduced, but much of the air simply recirculates throughout the building. When these systems are not working properly, pollutants and contaminants from office equipment, building materials, furnishings, carpeting, and smoke are trapped within the building. Two types of health problems have been identified in these types of buildings: **building-related illness** (DRI) and **sick-building syndrome** (SBS). (See *Outgassing, Baubiologie, Indoor pollution, Humidifier fever, Formaldehyde*)

heavy-metal pollution   Heavy metals are natural elements such as lead, mercury, cadmium, and nickel. They are mined from the earth and used in numerous manufacturing processes and countless products. Chromium and nickel are used to electroplate other metals to withstand corrosion and heat. Mercury is used as a fungicide in paint. Cadmium is used in plastics to stabilize colors. Lead is used in gasoline to boost octane, in marine and industrial paints as preservatives, and in car batteries. Thirty-five heavy metals pose a risk to human health and the environment.

Lead, the most prevalent heavy metal, has been phased out of most gasoline and was banned from most paints, but still exists in the paint found on millions of homes. Mercury is still allowed in most water-based paints and is known to cause symptoms and illnesses similar to **lead poisoning.**

Industry dumps its heavy metal wastes directly into bodies of water or into sewage systems, which make their way into our natural bodies of water. Shellfish in many regions are contaminated with heavy metals. Lead, leaking out of disposed-of car batteries into the groundwater, is one reason many landfills have been closed.

When materials containing heavy metals are incinerated, most of the metals go up the smoke stack, resulting in *air pollution.* Others placed into landfills, leach out into the soil and have been detected in groundwaters used for drinking.

Heavy metals can enter the body either through the air when we breathe, as *particulate matter,* on food or water, or they can be absorbed directly through the skin. They accumulate in the body, especially in the brain, liver, and kidneys, where they remain. If ingested, they can cause neurological or liver damage, or stomach cancer. Contact with these metals causes rashes or ulcerations while chronic inhalation can result in respiratory problems that range from coughing to lung cancer. (See *Mercury poisoning, Batteries*)

heavy-water nuclear reactor (HWR) Most of the nuclear power plants today are of the *light-water nuclear reactor* design, which uses common water (light water). Some reactors, however, are designed to use heavy water. Heavy water contains a different form (isotope) of hydrogen called deuterium, which is twice the weight of the hydrogen found in light water. Since heavy water works better than light water in controlling *nuclear fission,* the operating costs of an HWR are lower than those of an LWR. The basic design of an HWR is similar to that of an LWR, and they both have similar problems associated with them. (See *Nuclear power*)

herbage Refers to all the vegetation available to *grazing* organisms. Also called forage. (See *Consumer*)

herbicide Refers to a type of *pesticide* that kills *weeds* or all the vegetation in an area. Herbicides make up about 65 percent of all the *pesticides* used in the U.S. They are used to kill everything from a weed on your lawn (when they are commonly called weed killers) to all plant life along a proposed highway, railroad track, or under high-tension power lines. They are also used extensively on crops to rid unwanted vegetation from stealing water or sunlight from the crop to be harvested.

Herbicides are usually categorized according to how they kill and the method by which they enter the plant. Most herbicides kill by inhibiting some process such as photosynthesis, cell division, respiration, chlorophyll production, or root, shoot, or leaf growth. Some promote excessive growth by using a substance that mimics the plant's natural growth hormone, called auxin. This substance forces the plant to grow abnormally large, killing the plant since it cannot obtain enough nutrients to support its size. Herbicides that use synthetic hormones can be highly selective, and only affect the specific pest.

Herbicides enter a plant by one of three methods: 1) Contact herbicides kill on contact and include the well-known paraquat. 2) Systemic herbicides are absorbed and include *alar* and an ingredient in **Agent Orange;** and 3) Soil sterilants are placed in the soil, where it kills the microbes essential to the plant's growth. One such sterilant is trifluralin.

herbivore Animals that only eat plants (primary **consumers**) are called herbivores. Grazing animals such as cattle, sheep, rabbits, and grasshoppers are all examples of herbivores. Human herbivores are called vegetarians. (See *Food web, Energy pyramid*)

hermaphrodite An animal that contains both male and female reproductive organs or a plant that contains both types of organs in a single flower. The earthworm is a hermaphrodite. (See *Parthenogenesis*)

herpetology The study of reptiles and amphibians.

heterotrophs See *Consumers.*

high-market value farming counties Refers to U.S. counties that are in the top 20 percent of agricultural production for each state. More than half of these counties are inside or adjacent to metropolitan areas and are threatened by **urban sprawl** (expansion). Each year, about 2.5 million acres of agricultural land in the U.S. is converted to other uses in this expansion, with much of it being the most fertile and productive lands. (See *Farmland lost*)

high-temperature, gas-cooled nuclear reactor (HTGCR) Most nuclear power is generated by **light-water nuclear reactors (LWR),** but a few are either **heavy-water nuclear reactors (HWR)** or HTGCR. The main difference between these three is the substance used as a coolant. The HTGCR design uses helium gas, as opposed to regular (light) water in the LWR and a heavier form (isotope) of water in the HWR.

highwall Refers to the resulting damage caused by contour **strip mining** for coal when the land is not restored.

hinterland   An area far from civilization, unpopulated and un-developed. (See *Badlands*)

Holdridge Life Zone System   See *Zones of life.*

hormone weed killers   Refers to manmade, organic **her-bicides** that control the growth of weeds by producing an effect similar to the plant's natural growth hormones (called auxins).

host   A living organism that provides food or shelter for another organism. (See *Parasite, Symbiotic relationship*)

humidifier fever   A respiratory illness with symptoms much like the flu, caused by microorganisms that inhabit and flourish in humidifiers and air-conditioning units when not properly maintained and cleaned. (See *Baubiologie, Healthy homes*)

humus   Refers to a dark-brown soil-like substance formed by the partial decomposition of plants and animals, usually found mixed with the surface layer of **soil.** Humus supplies new plants with most of the **nutrients** needed for growth. It changes the texture of the soil and increases its ability to absorb water. (See *Soil organisms*)

hydrocarbons   Hydrocarbons are one of the five primary pollutants that cause **air pollution.** Hydrocarbons are released by the incomplete combustion of fossil fuels or by simple evaporation of fuels such as gasoline. Hydrocarbons by themselves are not a problem, but when they react with other primary pollutants, they form dangerous **secondary air pollutants.**

Positive crankcase ventilation (PCV) valves and gas cap air pollution control valves (APC), required on cars sold in the U.S., help reduce hydrocarbon emissions. Catalytic converters also help reduce hydrocarbons (along with other primary pollutants) spewing from the tailpipe. (See *Automobile fuel alternatives*)

hydroelectric power   Electricity produced by water movement has been used for decades. About 21 percent of the world's electricity and 10 percent of the U.S. electricity is generated by hydro power. The first hydroelectric plant was built on Niagara Falls in 1878. Hydropower can be generated by water falls, rushing rivers and streams, and manmade dams, all of which allow a controlled amount of water to pass through pipes that spin turbines—creating electricity. Enormous dams such as the Hoover (1,455 megawatts) and the Grand Coulee (6,180 megawatts) produce large quantities of power, but were environmental disasters and have become cost prohibitive to now build.

Most larger plants consist of a dam that backs up the water, raising the level. The released water falls into a turbine that generates electricity. Smaller plants do not necessarily require dams. They use a series of pipes with turbines inside which are turned by the current. This has less of a negative impact on the local *ecosystem.*

Many of the newer plants being built are small "mini-hydro" plants that generate less than 10 megawatts (MW) of power and even "micro-hydro" plants that create less than 1 MW of power. These small plants are often built in remote areas and create electricity for the immediate areas. About 1,500 such plants have been built in the U.S. China has built over 80,000 of these small hydro plants.

Hydro power is competitive, costing between three and six cents per *kwh.* They are also easy to control and can be turned on and off depending upon need. However, major changes in weather, such as drought, can reduce production. In the U.S., the 1988 drought caused a 25 percent drop in hydroelectric power for the year.

The Federal Energy Regulatory Commission has identified about 7,000 sites within the U.S. that are feasible for hydro power development, capable of producing over 150,000 MW of power. So far, about 2,000 of these have already been developed and are creating about 70,000 MW of power. One hundred and seventy-four more are under development. However, many of the remaining sites have been put off limits for future hydro development by the 1968 National Wild and Scenic Rivers Act, which prohibits development on virgin rivers and streams. This act protects about 40 percent of those sites identified as feasible for a hydro plant, limiting the potential for hydro power to help supplant fossil fuels in the future.

Hydroelectric power creates virtually no pollution problems. Small-scale projects cause little harm to the environment, but larger projects are environmentally destructive.

Moon power, wave power, ocean thermal energy conversion, and *solar ponds* also use water power in various forms. (See *James Bay Project*)

hydrogen gas   Hydrogen gas is an experimental source of energy that theoretically has the potential to provide all our energy needs. It burns cleanly, producing only water vapor and has 2.5 times the energy content of gasoline. The major problem with this fuel is its lack of availability. It is not abundant in nature and must be produced. If more energy is needed to create a fuel than can be supplied by that fuel, it is impractical as an energy source. This means efficient methods of producing hydrogen fuel must be found to make it useful.

The best hope is an efficient type of solar cell that breaks water down into hydrogen and oxygen molecules. If this technology can be perfected, it could relieve our dependence on **fossil fuels** and reduce many of the environmental problems associated with them. Japan and Germany are heavily involved in this research. (See *Alternative energy*)

hydrologic cycle   See *Water cycle.*

hydroponic aquaculture   Refers to growing crops and fish together in closed, circulating systems, in which fish are raised for food and hydroponic vegetables or other plants are grown in the water on the waste from the fish. Some companies are using hydroponic aquaculture to grow fish and vegetables to be sold to organic food supermarkets. (See *Hydroponics*)

hydroponics   Refers to growing plants in an artificial liquid environment, using an artificial nutrient solution, and providing artificial support for the plant.

hydrosphere   All forms of water on our planet constitute the hydrosphere. The majority of the hydrosphere consists of **surface waters,** which covers 74 percent of the earth's surface. This includes oceans, freshwater and saline lakes and rivers, and ice caps and glaciers. **Groundwater** and moisture in the soil is also considered part of the hydrosphere.

Human intervention is affecting the quality of water in many ways, including **acid rain** and **groundwater contamination.** There has even been an increase in the amount of lead found in layers of ice in the ice caps. (See *Water use, Water cycle, Aquatic ecosystems, Biosphere*)

hyperaccumulators Refers to plants or animals that thrive on contaminated soils. Some plants can absorb high concentrations of contaminants from the soil, thereby decontaminating it. Research is currently being done with plants that hyperaccumulate **heavy metals.** After the plants have accumulated the metals, they can be disposed of as a **hazardous waste** or possibly processed to remove the heavy metals and recycle them for further use.

Plants such as common ragweed and hemp dogbane remove lead from soil, while others have been found to remove zinc, cadmium, and nickel. Once the plants have been harvested with high concentrations of the metals, they act like a bio-ore, which is then processed or "smelted" like mineral ores for the metal content. Contaminated soils within **Superfund** sites might be decontaminated in this way in the future. (See *Bioremediation, Hazardous waste remediation, Jimsonweed*)

ice ages Gradual climate changes, lasting thousands of years, resulted in dramatic transformations of our planet's environment. Over the past 750,000 years, these changes resulted in eight great ice ages in which huge sheets of ice moved down from the polar cap, blanketing large portions of North America, Europe, and parts of Asia. Each ice age lasted up to 100,000 years at which time, gradual warming allowed the ice caps to recede.

The periods between the ice ages, called interglacial periods, only lasted about 10,000 to 12,000 years each. The last ice age drew to a close about 10,000 years ago. The difference in temperature between the interglacial period (we are now in) and the last ice age is only about 5°C (9°F). (See *Global warming, Albedo*)

ichthyology The study of fishes. (See *Aquatic ecosystems*)

igneous rock Rock formed from cooled, hardened molten magma.

immature Refers to any stage of development in an organism prior to reaching sexual maturity.

immigration Immigration is one of two forms of ***migration*** across international borders. Immigration refers to the movement of people into a country, and emigration refers to movement out of a country. Immigration and emigration, along with ***fertility*** and ***mortality,*** dictate changes in the size of a nation's ***population.*** (See *Immigration, replacement level*)

immigration, replacement–level *Immigration* is the flow of people from one country into another. About 700,000 people legally immigrate into the U.S. each year and roughly 200,-000 immigrate illegally. This contributes a significant portion of the total annual population growth in the U.S. Replacement-level immigration refers to limiting the number of immigrants into the U.S. to the same number as those who emigrate out of the country annually, which is about 130,000 people. (See *Doubling-time in human populations*)

impact equation A simple equation introduced by **Paul Ehrlich** that demonstrates the relationship between people and their impact on the environment: impact per person × number of people = total environmental impact. (See *Populations, Paul Ehrlich, Doubling-time in human populations*)

imperfect flower A unisex *flower,* containing only the male (pistil) or female (stamen) reproductive organs. For example, the scrub oak produces imperfect flowers.

In Business Aptly called by the publisher *The Magazine for Environmental Entrepreneuring,* this bimonthly publication is for anyone interested in a business related to the environment. Available on newsstands and by subscription, it contains plenty of good advice and information about different industries and markets. Tidbits of information and full-length articles such as "Organic Foods Find Growing Shelf Space" are informative for both the businessperson and the consumer as well. Call (215) 967-4135. (See *Ecomagazines, Ecopreneur*)

in-stream water use In-stream water use is one of four ways people use water. In-stream refers to using a flow of water for human activities and includes **hydroelectric power,** navigation, and recreation, including boating, swimming, and fishing. Water recreation requires clean, unpolluted waters. To sustain this, the public using the water must be held accountable to keep the water clean.

Navigable waterways often require dredging or widening, which often causes environmental damage to the **aquatic ecosystem.** Today most projects that affect waterways require environmental impact statements that analyze the potential damage. (See *Water use*)

incineration Incineration is one of four methods of **municipal solid waste disposal.** Incinerators burn garbage to reduce its volume. Some use the heat to generate electricity in **waste-to-energy power plants.** There are over 100 such facilities in the U.S., with many more planned. The major advantage of incineration is a reduction in volume, meaning there is less that need be placed into landfills. Burning solid waste reduces its volume between 60 percent and 90 percent. (See *Incineration problems, Biomass energy*)

incineration problems *Incineration* is one method of disposing of ***municipal solid waste.*** The major advantage of incineration is that it reduces the volume of solid waste, but there are also problems associated with its use. Where does the incinerated waste go? Most goes up the smokestack into the air, causing ***air pollution.*** Air emissions from incinerators are called fly ash, and they often contain hazardous substances such as ***dioxin, heavy metals,*** and acid gases. Pollution control devices such as scrubbers and electrostatic precipitators reduce but do not eliminate these pollutants before they leave the smokestack.

The solids that remain after burning, called bottom ash, must still be placed in landfills and may contain concentrated toxic substances that didn't go up the smokestack. Bottom ash from some incinerators must be handled as ***hazardous waste*** and cannot simply be dumped into landfills.

Opposition to new incineration facilities because of the reasons mentioned above have stopped the production of many planned incinerators because of the ***NIMBY (not in my backyard) syndrome.*** Alternatives to incinerators and landfills are ***recycling*** and ***source reduction.***

Index of Sustainable Economic Welfare (ISEW)

The ISEW is an alternative method of evaluating the overall health of an economy and the inhabitant's well-being. For 50 years, the economic health of a country has been measured by the gross national product (GNP). The GNP measures the total output of goods produced and services provided, but it does not take into account the depletion of natural resources or the negative impact it might have on the environment or its people.

For example, irrigating fields raises productivity by producing crops, but reduces the water supply for the next generation. Burning coal raises productivity but pollutes the air, degrading the quality of life for those in the region.

The Index of Sustainable Economic Welfare does incorporate environmental degradation into its calculations. It takes into account the depletion of renewable resources, the loss of farmland and wetlands, and the cost of water and air pollution. It also includes long-term factors, such as the impact of a product that damages the ozone layer or increases global warming.

The ISEW has shown a gradual decline (about 12 percent) from 1976 to the present. Unfortunately, this indicator is currently calculated only for the U.S. Many countries don't even have the data available to calculate the ISEW. The ISEW may prove to be a transitional indicator, helping the world find better ways to measure the health of economies and the well-being of populations now and for the future. (See *Economy vs. environment, Sustainable agriculture*)

indigenous Refers to organisms (including humans), that originally lived in an area, as opposed to being introduced from elsewhere. For example, Native American Indians are indigenous to North America.

indoor pollution Most people worry about outdoor *air pollution,* and rightly so. Far too few people, however, think about the quality of air indoors, where they spend 90 percent of their time. *Sick-building syndrome* and *building-related illness* are examples of how indoor pollution affects our health. *Healthy homes* and the study of *baubiologie* are new fields that try to address and rectify problems that exist, or prevent them from happening.

Two factors play a major role in the increase in indoor pollution: the increased use of chemicals and synthetic materials in the construction of buildings and the move toward energy efficiency by building tightly sealed buildings.

Substances such as *formaldehyde* are released from building products into the air by the process of *outgassing. Asbestos* particles, along with numerous other harmful substances, are found in many building materials and furnishings. New carpeting alone emits the toxins: formaldehyde, ethyl benzene, toluene, xylene, and other *volatile organic compounds (VOCs).*

The fact that buildings are often sealed shut concentrates these and other substances. *Pesticides, heavy metals* from road dust, naturally produced *radon,* and *pathogenic* microbes can accumulate indoors and thrive in *heating, ventilation, and air-conditioning systems (HVAC).*

The biggest indoor air problem, however, is smoke from cigarettes, including *passive smoke* breathed in by nonsmokers. It is estimated that 350,000 people in the U.S. die each year from smoking-related illnesses.

Industrial melanism Refers to an increase in the number of dark-pigmented individuals (such as moths) in response to a habitat becoming blackened due to industrial pollution. (See *Natural selection, Speciation*)

industrial smog   Refers to smog containing high levels of **sulfur dioxide** and sulfuric acid produced by burning coal containing sulfur impurities. Pollution control devices eliminate enough of these contaminants to make industrial smog rare; also called gray smog.

industrial water use   Industrial use is one of four ways people utilize water. The majority of industrial water is used for cooling. Most electricity-producing power plants use water as a coolant. Paper mills and many manufacturing processes require vast amounts of water. Recycling the water once used is practiced by some manufacturers and dramatically reduces the volume needed. Water used during manufacturing processes often becomes contaminated, causing **industrial water pollution.** (See *Thermal water pollution, Water use for human consumption*)

industrial water pollution   *Industrial water use* causes water pollution since its final destination, just like **domestic water,** is streams, rivers, the ocean, or groundwater. Some industry waste is combined with domestic wastewater, but much of it is handled separately. New industrial facilities are required to treat their wastewater before releasing it, but older plants often dump the wastewater directly into a river or stream.

Industrial wastewater may contain oil-based products, **heavy metals,** acids, salts, or organic substances. In some cases these substances are highly toxic and should be handled as **toxic wastes. Thermal water pollution** is another form of industrial water pollution, caused when water is used for cooling. The water becomes hot and must be released. Discharging this hot water into natural bodies of water changes the water temperature and impacts the **aquatic ecosystem.**

infectious disease   Refers to any disease transmitted without physical contact. (See *Contagious disease*)

Inform   Inform, established in 1974, is an **environmental organization** that concentrates on: chemical hazards prevention, **solid waste management,** urban air quality, and land and water conservation. Its purpose is to educate and provide practical measures to reduce pollution and conserve resources. Inform publishes many excellent materials that inform the public about these issues, and it helped shape the first federal legislation on waste prevention introduced in Congress. Write to Inform, Inc., 381 Park Avenue South, New York, NY 10016.

inholdings   Inholdings are parcels of private land surrounded by public lands. (See *Land acquisition, wilderness; Nature conservancy*)

inland wetlands **Wetlands** are divided into two types: **coastal wetlands** and inland wetlands. About 95 percent of all remaining wetlands in the U.S. are inland wetlands. Wetlands play a vital role on our planet. They provide a habitat for numerous fish, waterfowl, and wildlife. They help control flooding by storing vast quantities of rainwater, which is then used to recharge (replenish) **groundwater** that supplies much of our domestic water. Wetlands filter out sediment and dilute pollutants from incoming streams, which would otherwise end up in our water supply. Crops such as rice, blueberries, and cranberries are cultivated in these areas.

Over 50 percent of the wetlands in the continental U.S. have already been destroyed. Of the remaining inland wetlands, only about 8 percent is under federal protection. Many local laws are insufficient to prevent individuals from developing this land.

inorganic matter   Refers to substances that are not alive and do not come from decomposed organisms. (See *Organic matter*)

insect sterilization   Insect sterilization is an innovative way of controlling insect pests. It is an alternative to **pesticides** and may be used as part of an **integrated pest management** program. Insect sterilization involves mass-rearing a pest insect and sterilizing the males with either chemicals or radiation. Once sterilized they are released at the appropriate time in the infested area to mate with unsuspecting females. Since they are competing with wild, nonsterilized males, they must far outnumber the virile males by 10 to 1. The release is usually done multiple times and works only on species that mate only once.

The screwworm fly, which is an often fatal parasite on cattle, goats, and deer, has been virtually eliminated in many parts of the U.S., Mexico, and Central America using this method. (See *Sex attractants, Biological control, Natural insecticides*)

insecticides   Refers to any substance that kills **insects.** The vast majority are synthetically produced and can be placed into one of four categories: **chlorinated hydrocarbons, organophos-phates, carbamates,** and **pyrethroids.** An alternative to these synthetic insecticides is **natural insecticides** that use substances found in nature. (See *Biological control, Biological pesticides, Insect sterilization*)

insectivore   Refers to an organism that feeds on **insects.** Insectivores include numerous birds, fish, reptiles, amphibians, and some mammals such as shrews and anteaters, and many **predatory** insects. Also, a few plants such as the pitcher plant and the well-known Venus's-flytrap can ingest insects.

insects   There are about 1.4 million identified species (plants and animals) on our planet. Roughly 1 million of them are insects, and about two-thirds of these are **beetles.** The typical insect body is divided into three regions (head, thorax, and abdomen) and the thorax typically contains three pairs of legs. (See *Pesticides*)

insulation   See *Thermal insulation.*

integrated pest management (IPM)   Integrated pest management is an alternative to using synthetic **insecticides** exclusively. IPM uses a variety of techniques in a carefully planned program designed to reduce our dependence on **pesticides.** IPM uses other techniques and technologies available to us, many of which are at the forefront of biotechnology. These techniques include **insect sterilization, sex attractants, resistant crops,** cultural or **organic farming** techniques, **natural insecticides, biological control,** and the selective use of synthetic pesticides.

IPM is not more widely used due to the lack of coordinated programs sponsored by local, state, and the federal government, and the needed funding. The continued use of dangerous pesticides as our primary form of pest control, in spite of the availability of alternative technologies that exist, is a management and education problem that should be resolved. (See *Pesticide dangers, Pesticide residues*)

intelligent vehicle–highway systems (IVHS)
Millions of research dollars are spent each year in the U.S. on ways to create "intelligent" highways. Optimistically, these technologies hope to reduce the average commute in crowded areas by 50 percent. Some of the simpler aspects of this technology are already in

operation in some areas. For example, some roads monitor traffic flow with magnet-induction loop detectors embedded in roads at half-mile intervals. The traffic data is collected by roadside devices, which pass the information along to a computer. The computer controls traffic lights and message boards in an effort to relieve congestion.

More advanced plans for IVHS include sending traveler advice directly into navigation receivers in cars and someday actually controlling a car's speed, braking, and possibly even steering. (See *Mass transit, Ecocities, Automobile fuel alternatives*)

intercrop Refers to growing two or more crops together in the same plot of land. (See *Alley cropping, Soil conservation*)

interglacial period A warm period that exists between two **ice ages.**

International Joint Commission (IJC) The IJC is composed of three members from the U.S. and three from Canada. This organization carries out environmental studies and makes recommendations to policymakers in both the U.S. and Canada about bodies of water that are shared by both countries, such as the Great Lakes.

invasion The mass movement of a population into a new area, often displacing an existing population. (See *Migration*)

invertebrates Refers to those animals that don't have backbones. Support is provided by other means, such as an exoskeleton (as in **insects**) or by floating in water (such as with jellyfish). (See *Coral reefs*)

irradiation Refers to exposure to **radiation.** (See *Nuclear reactor safety and dangers, Food irradiation*)

irrigation Human water use is divided into four categories: **domestic water use, industrial water use, in-stream water use,** and agricultural irrigation. Irrigation involves transporting or diverting water from the source to regions needing the water. For example, 35 million acres of crops grown in arid regions of the western U.S. are irrigated with water piped from hundreds of miles away.

Irrigation uses 70 percent of the water used by humans worldwide. Since two-thirds of the water used in irrigation evaporates or runs off, and never gets to the plants in need, new methods of irrigation have been developed. (See *Drip irrigation, Water pollution*)

ivory trade   During the past decade, the number of elephants in Africa has been reduced by half, principally due to the international ivory trade. The **Convention on International Trade in Endangered Species (CITES)** has tried to control the ivory trade since 1976, when almost 90,000 elephants were slaughtered each year for their ivory tusks. Since 1986, the organization has tried to completely ban the ivory trade. It has met with moderate success, but is constantly battling **poachers.** Some members of the organization feel that the ban should be lifted since the elephant population has stabilized, but most feel this would reinitiate a significant drop in population. (See *Endangered species, Endangered Species Act*)

Izaak Walton League of America   This organization was established in 1922 by a small group of concerned anglers. It has grown to 53,000 members dedicated to protecting the nation's soil, air, woods, waters, and wildlife. They are involved in protecting areas for recreation and improving recreation-landowner relations. They publish several newsletters and a quarterly magazine, which discusses articles on conservation and recreation. Membership is 20 dollars. Write to 1401 Wilson Blvd., Level B, Arlington, VA 22209.

J-curve   Theoretically, the *population* of any organism could continue to grow until it took over the entire planet. For this to occur, there would have to be unlimited resources such as food and shelter, and ideal conditions such as temperature and quantity of available water. If such a curve was plotted on a graph, it would look something like the letter "J," starting off with a small population in the beginning and ending with a large population at the end.

Since resources are not limited and conditions are not always perfect, this ideal curve doesn't usually occur in nature. Once the population reaches the limits of what the environment can support (called the *carrying capacity*), growth levels off, creating a curve looking more like the letter "S," which is experienced by virtually all organisms, with the exception of humans.

The human population has yet to level off and is still on the upward swing of the J-curve. Technology has so far forced our planet's "human" carrying capacity to support our numbers, but in many regions of the world, it appears to be pushed to the limit. For example, the infant *mortality* rate is four times greater in the *less developed countries (LDCs)* than in the *more developed countries (MDCs).* Only time can tell when the earth's carrying capacity for the human population will be reached and exactly how it will come about. (See *Doubling-time in human populations, Thomas Malthus, Population stability, Population dynamics*)

jactitation   Some plants disperse their seeds by a jerking motion, which tosses the seeds from the fruit. This process is called jactitation.

James Bay Power Project   The purpose of the James Bay Power Project is to harness energy in the rivers flowing through northeastern Quebec. Phase I of this project has already been completed, but a massive Phase II is getting underway. The water would be collected in huge reservoirs. Hydro-Quebec, the company responsible for this project, asserts that dikes, powerhouses, roads, and transmission lines installed in this wilderness can be accomplished with little environmental harm. Proponents say that power from James Bay would reduce the number of coal- and oil-fueled power plants and nuclear power plants needed in the future, which would cause even more harm to the environment than the hydro project.

The habitats of many species would be destroyed by the project, including moose, caribou, beaver, and numerous fish and waterfowl. The breeding grounds of millions of migratory birds would also be damaged. The toll the project might have on the larger ecosystem of the entire region is unknown. (See *Hydroelectric power*)

jimsonweed   A plant recently discovered to neutralize soils contaminated by radioactivity. With further research, Jimsonweed and similar types of plants may prove useful in helping resolve our **nuclear waste disposal** problem. (See *Hyperaccumulators, Phytoremediation*)

joule   The quantity of energy that can be generated from a fuel such as oil or gas is measured in units called joules. One joule equals the energy necessary to raise one kilogram of weight, 10 centimeters in height. Large quantities of energy are measured in megajoules, abbreviated MJ (one million joules), and gigajoules, GJ (one billion joules).

jungle   Refers to **habitats** characterized with high precipitation, usually found in tropical regions. (See *Tropical rain forest, Biome*)

junk mail   This is unsolicited, mass-distributed marketing materials sent through the mail. It is considered by many environmentalists an incredible waste of natural resources, since millions of trees are required annually to produce the volume of junk mail generated. The typical family of four receives almost 1,000 pieces of junk mail per year. Estimates state that about 34 million trees are used annually just to create junk mail, producing about 2 million tons of trash that must be dumped into **landfills** or **incineration** plants. To stop receiving unsolicited mail-order catalogues, write to Direct Marketing Association, Mail Preference Service, P.O. Box 9008, Farmingdale, NY 11735, and indicate which catalogues you no longer want to receive. (See *Paper recycling*)

K-strategists   Organisms that are usually large, have relatively long lives, and produce few offspring are called K-strategists. These organisms invest a great deal of energy in assuring the survival of the few offspring they have. Most large mammals, such as horses, deer, and humans, are K-strategists. The population size of these organisms is usually limited by density-dependent **limiting factors.** This means the population grows until the density of the population limits any further increase.

For example, the population of **predatory** birds such as the hawk continues to increase until there are too many hawks and too few snakes, mice, and other prey available for the young hawks to survive. Food becomes the limiting factor, and the hawk population stops increasing. K-strategists have populations that stabilize at the **carrying capacity.** (See *R-strategists, J-curve*)

karst terrain   When water makes its way into the soil, it sometimes passes through regions containing limestone or gypsum. The water mixes with acids in the soil, which dissolves, cracks, and otherwise breaks down the rock. The deteriorated substrate is called karst terrain and results in many underground caves and surface **sinkholes.** (See *Groundwater, Aquifers*)

keystone species   If a single species is of critical importance to the stability of an **ecosystem**, it is called a keystone species. Removal of this species could result in the collapse of the existing ecosystem. The alligator in the southeastern U.S. is an example of a keystone species. Alligators dig holes that accumulate water, which later becomes a habitat for many forms of aquatic life. They also build nesting mounds that turn into nests for herons and egrets. Alligators eat large amounts of predatory fish; this stabilizes the **populations** of the prey, such as bass.

When the alligators were almost hunted into extinction in the 1950s and 1960s, the entire ecosystem in the region was altered. The aquatic organisms lost their habitat, birds lost their nesting grounds, and populations of fish dramatically shifted since the predatory fish were no longer kept in check by the alligator. (See *Food webs, Population dynamics*)

kilowatt (kW)   A kilowatt is used to measure the electrical generating capabilities of a power plant. A small windmill might generate two kilowatts of power, while a coal-burning plant may generate thousands of kilowatts. (See *Watt, Kilowatt-hour, Btu, Energy sources*)

kilowatt–hour (kWh)   Refers to one kilowatt of electricity supplied for one hour. An average household consumes between 500 and 1,000 kWh per month.

kingdom   A kingdom is the highest category in the system used to classify life on earth. Some classification systems have as few as two kingdoms while others have as many as five. The most accepted classification today divides organism into five kingdoms: 1) Animalia (animals); 2) Plantae (most *algae* and all plants); 3) *Fungi* (all true fungi); 4) Protista (protozoa and slime molds); and 5) Monera (*bacteria* and blue-green algae).

  The kingdoms are divided into phyla (phylum is singular), which are in turn divided into classes. For example, humans belong to the kingdom Animalia, phylum Chordata and class Mammalia, while insects belong to the kingdom Animalia, phylum Arthropoda and class Insecta. (See *Species, Natural selection*)

kleptoparasitism   Refers to a form of parasitism in which an individual of one species steals food from another species to feed its own young.

Kricket Krap   Kricket Krap is a unique, fun, and entrepreneurial alternative to chemical *fertilizers* for your house and garden plants. It is the accumulation of droppings (fecula) from about 2 billion crickets that are raised as fish bait in Augusta, Georgia. Call (404) 722-0661. (See *Zoo Doo, Organic farming, Ecopreneur, In business*)

kymatology   The study of waves.

land acquisition, wilderness There are two major forces involved in the acquisition of wilderness areas: the federal government and **nongovernmental organizations (NGOs).** The government procures lands with the Land and Water Conservation Fund, Migratory Conservation Fund, and the North American Wetlands Conservation Fund. The Land and Water Conservation Fund is funded by some of the royalties collected from oil- and gas drilling on federal lands. Most of the lands acquired are near or within the borders of existing national parks or forests. Over the past 27 years, 35,000 projects have benefited from over $3 billion in funds approved by Congress.

The Migratory Bird Conservation Fund was established to restore and enhance migratory bird habitats, with some of its funds coming from the sale of duck stamps and others from entrance fees at refuges and firearms import duties. About $30 million is invested in wetlands each year. The Wetlands Conservation Fund passed in 1989 by Congress authorized the purchase, management, and restoration of **wetlands** in the U.S., Canada, and Mexico. Determining exactly what is a wetland and saving these areas in the past few years has been more of a political football game than a scientific endeavor.

NGOs, such as the **Nature Conservancy,** have had great success in saving numerous wilderness areas.

landfill About 80 percent of all U.S. **municipal solid wastes** are dumped into landfills. Today's modern landfills, also called sanitary landfills, are quite sophisticated. They contain liners to collect fluids (called **leachate**), which is pumped away for treatment, and gas collection systems to remove **methane** gas as it is produced. This gas can be used as a fuel to generate electricity. Monitoring devices surrounding the site test if the **groundwater** beneath is becoming contaminated or methane gas is escaping. As waste is dumped on the site, it is covered with a layer of clay or other material at regular intervals.

Contrary to popular belief, the material in a landfill is not expected to decompose to any great degree, which is why "fill" is part of the name. Water and oxygen are required for rapid decomposition to occur. Since most landfills are kept relatively dry (to prevent leachate from contaminating the groundwater) and little oxygen enters deep in the pile of waste, the contents aren't allowed to decompose.

Once a landfill is "full," it is graded, capped, and usually turned into some form of recreational area such as a park, golf course, or athletic field. Due to many problems associated with landfills, the number of them is dwindling throughout the world, and alternatives are needed. (See *Landfill problems*)

landfill problems   In 1979, there were 18,500 **landfills** operating in the U.S. In 1988, there were only 6,500, and it is estimated that by the year 2000, there will be half that number. Some people think the decline is due to a lack of adequate space for landfills, but space is simply not the problem. There are two reasons for the worldwide decline in the number of landfills. First, almost all older landfills and probably over 25 percent of the newer landfills have been found to leak contaminants into the **groundwater** supply, damaging our most important resource, water.

The second reason is the **NIMBY syndrome** (Not In My Backyard). Even though land exists for more landfills, they must be relatively close to the municipality producing the garbage, and people obviously don't want garbage and contaminated water in their community. These two problems, combined with the fact that most existing landfills are filling-up, indicate that landfills will not remain our primary **municipal solid-waste** disposal method in the future. The U.S. is not alone in moving away from landfills; Japan, Sweden, and Switzerland place less than 15 percent of their waste into landfills. **Recycling** and **incineration** are alternative methods of solid waste disposal. (See *Hazardous waste, Toxic waste, Nuclear waste disposal, Source reduction*)

laser printer ozone   Laser printers, used with personal computers, can emit excessive amounts of ozone if their filters are not routinely replaced. Ground-level **ozone** can cause respiratory ailments, nausea, and headaches. Dust clogs the filters and keeps them from working correctly. Hewlett-Packard, which has about three million laser printers in use, advocates that the ozone filters should be changed after about every 50,000 pages of print. (See *Computer-produced pollution, Indoor pollution, Copier toner cartridges, recycled*)

law and the environment   As a wave of environmental litigation begins to pass through the judicial system, prosecutors, defense attorneys, and judges are trying to increase their level of **environmental literacy** about topics such as **wetlands** protection, **air pollution, hazardous waste disposal,** and **risk assessment.** Special seminars and classes have been available for lawyers for quite some time on these subjects, but only recently have judges been targeted. The Flaschner Judicial Institute in Boston, Massachusetts, was one of the first to offer an environmental seminar expressly for judges. (See *SLAPP, SLAPP BACK*)

lawn mower pollution   In one hour of running time, a gasoline-driven lawn mower produces the same amount of **hydrocarbons,** a primary component of **air pollution,** as does a car driven for the same period of time. Although much smaller and with less horsepower, the lawn mower is far less sophisticated in design and less likely to be tuned than a car. California is beginning to regulate the design of lawn mower engines, and other states are expected to follow suit. Many manufacturers are making changes in advance of pending regulations.

Electric lawn mowers are a viable alternative, except for the inconvenient cord that comes with them. New technologies are being tested that include a rechargeable electric mower that can mow a quarter of an acre before needing recharging. (See *Automobile alternative fuels, Alternative energy*)

LD50   LD50 is the commonly used unit to measure the toxicity of a substance. LD stands for Lethal Dose, and 50 refers to 50 percent. LD50 measures the dose required to kill 50 percent of the individuals in a sample population exposed to the substance. It is measured by body weight in kilograms. For example, the substance that causes botulism (a food poisoning) has an LD50 of .0014 milligrams. This means a group of animals weighing 100 kilograms each that have .14 milligrams of the substance in their bodies (one milligram for each kilogram of body weight) would result in the death of 50 percent of all the animals. (See *Acute toxicity, Chronic toxicity, Toxic waste, Pesticide residues*)

LDPE   LDPE is an acronym for low-density polyethylene, which is a type of **plastic** commonly used to manufacture films and plastic bags. It makes up about 30 percent of all the plastics produced in the U.S., but is rarely recycled. These products are often stamped with the number "4" surrounded by a triangle formed of arrows. (See *Plastic recycling, Plastic pollution*)

leachate   As precipitation falls on a **landfill,** water gradually passes through (percolates), becoming a soup of decomposing waste and microbes, called leachate. It often contains a variety of hazardous substances, including **heavy metals** and organic compounds. Regulations call for newer landfills to be lined to catch the leachate so it can be treated before being released into the environment. However, older landfills have no liners and even those with liners have been found to leak leachate, which eventually contaminates the **groundwater.**

leaching   As water flows over or through **soil** or rock, it removes chemicals and carries them elsewhere. This process is called leaching. For example, many of the nutrients found in topsoil leach down to the lower levels of the soil. (See *Soil profile*)

lead poisoning   Refers to the exposure to lead. This can occur by the inhalation of fumes containing lead or the ingestion of particles containing lead, such as from lead paint. Leaded gasoline, along with numerous other products, has resulted in lead contamination permeating the environment. Lead poisoning can cause acute illness, such as vomiting, or chronic illnesses, such as permanent damage to the nervous system, as lead accumulates in the body over long periods.

Lead is being phased out of many products such as gasoline and paints, but is still commonly found in many other products, including some unusual places most people wouldn't expect. Two examples are given below.

Avoid ingesting lead by washing off the lip of a wine bottle where the foil cap comes in contact with the bottle. The wine picks up the lead (originally in the foil) as it flows over the lip. If you recycle plastic food packaging such as bread wrappers, don't turn the plastic bag inside out, since lead in the painted labels can be released and absorbed into the bread. (See *Heavy-metal pollution, Mercury poisoning*)

leaf litter   Plants and plant parts that have recently fallen to the ground and are only partially decomposed are called leaf litter. Leaf litter often forms the surface layer in many **habitats.** (See *Soil organisms, Soil horizon, Duff*)

League of Conservation Voters (LCV) The LCV helps elect pro-environment candidates to office by endorsing and providing financial support to those with good environmental track records. They also have an excellent program to make the public aware of **greenscam.** This information is provided in the annual National Environmental Scorecard. With its goal of changing the balance of power in the U.S. Congress to reflect the pro-environment sentiment of the electorate, the LCV is one of the most important of all **environmental organizations.** The annual membership is 25 dollars. Write to 1707 L Street NW, Suite 550, Washington DC 20036.

lentic Refers to standing bodies of water such as ponds and lakes. (See *Standing water habitats*)

Leopold, Aldo Aldo Leopold, a conservationist, writer, and philosopher, is best known for his environmental perspective in the essay "The Land Ethic" from his book entitled *A Sand County Almanac and Sketches Here and There.* Leopold founded the field of game management. While working for the **Forest Service** during the 1920s, he helped develop the wilderness policy and advocated regulating hunting to maintain a proper balance of wildlife in a habitat. (See *Philosophers, environmental; Ethics, environmental*)

less developed countries (LDC) Defined by the United Nations as a country with low to moderate industrialization and low to moderate gross national product. There are 150 LDCs, with most located in Africa, Asia, and Latin America. Most have less favorable climates and soils compared with the **more developed countries (MDC).** About 77 percent of the world population lives in LDCs. The annual population growth rate of these countries is 2.1 percent (compared with 0.5 percent for the MDCs). Living conditions in many of these countries is often at a subsistence level. For example, almost half of the population in LDCs have unsafe drinking water (compared with almost none in the MDCs). (See *Standard of living*)

levels of biological organization, study of
Ecological studies often focus on a certain level of biological
organization. Studying individuals or a single *species* in an *ecosystem* is called *autecology.* Studying one *population* or a
few interacting populations is *population ecology.* Studying all
the populations within an *environment* is *synecology. Systems ecology* (or ecosystems science) studies all the relationships
between all the organisms and nonliving factors found in an environment.

lichens These are organisms that consist of *algae* and *fungus*
in a *symbiotic relationship.* These organisms are capable of
living in the harshest of environments, including on bare rock. They
are often the first organisms to appear in a *pioneer community.*
The algae photosynthesize and pass nutrients along to the fungi.

life zones Using life zones is one of many methods to divide life
on earth into major geographical regions. The method was proposed at the turn of the century and consists of transcontinental
belts running east and west, based on temperature. A more recent
variation of life zones is the Holdridge Life Zone System, which is
more complex and takes into account many more variables than just
temperature. (See *Zones of life*)

light–rail Refers to *mass-transit* commuter trains, trolleys,
and street cars used as an alternative to the automobile. (See *Ecocities, Automobile fuel alternatives, Bicycles, Urban sprawl*)

light–water nuclear reactor (LWR) Almost all existing *nuclear power* plants are of the LWR design. (Light water
refers to common tap water.) These reactors use about 40,000 radioactive fuel rods, which are surrounded with water. The water acts
as a coolant and helps control the reaction. A chain reaction of
nuclear fission is started, and the energy released converts the
water to steam, which drives turbines to generate electricity.

After three to four years, the fuel rods, which contain uranium,
are no longer capable of supporting nuclear fission and must be
removed from the reactor and disposed of. These rods remain highly
radioactive and deadly to all forms of life, so they must be transported and stored in a *radioactive waste disposal* facility until
they are no longer harmful. This takes about 240,000 years, but a

mere 10,000 years if the rods are first processed for disposal. An alternative method of disposing of the radioactive fuel is to send the fuel to reprocessing plants, which recycles the fuel. Since reprocessing the fuel is more costly than the original processing, only a few of these plants exist, none of which are in the U.S. (See *Heavy-water nuclear reactors*)

Lighthawk Lighthawk is an organization dedicated to protecting the environment by flying decision-makers, media representatives, and grassroots activists over and into endangered lands. This provides them with the firsthand experience they need to take action. Some of its accomplishments include helping to cap the largest single source of arsenic air pollution in the United States—an aging copper smelter in southern Arizona—and creating the 92,000-acre Bladen National Park/Nature Reserve in Belize, Central America. Write to P.O. Box 8163, Santa Fe, NM 87504. (See *Environmentalist, Environmental organizations*)

limiting factor Refers to that factor in an organism's environment that is furthest from the optimum and therefore limits the organism's chances for survival. Limiting factors might be **biotic,** such as insufficient food (plants to eat or prey to be captured), or they might be **abiotic,** such as not enough sunlight, water, or phosphorus for plants to grow. (See *K-strategists, R-strategists, Carrying capacity, Tolerance range*)

limnetic region The limnetic region of a lake receives sunlight but is too deep for vegetation to take root. This region often contains abundant **plankton.** (See *Standing water habitats*)

limnology Refers to the study of freshwater organisms and their habitats. (See *Freshwater ecosystems*)

liquid metal fast breeder reactor (LMFBR) See *Breeder reactor.*

liquified natural gas (LNG) See *Natural gas extraction.*

liquified petroleum gas (LPG) See *Natural gas extraction.*

littoral zone   This refers to the shoreline between the high and low tide marks and the immediate area affected by the tides. (See *Neritic zone*)

loam   Refers to a type of soil consisting of 40% sand, 40% silt, and 20% clay. Loam is considered the ideal **soil type** for most crops since it has excellent aeration and drainage properties. (See *Soil organisms*)

lotic   Refers to **running water habitats** such as rivers and streams.

Love Canal   Love Canal symbolizes the problems associated with **hazardous waste disposal.** For about ten years, beginning in the late 1940s, Hooker Chemicals and Plastics Corporation dumped steel drums containing about 22,000 tons of **toxic wastes** into an old canal (named after its builder William Love). In 1953 the company placed topsoil over the site containing the drums and gave the property to the Niagara Falls school district. A school, recreational fields, and almost 1,000 homes were built on the site over the next few years.

Beginning in 1976, residents began to notice odd smells and found that children playing around the canal often received chemical burns. Concerned citizens, including **Lois Gibbs,** performed informal health studies that revealed high incidences of many types of disorders. The drums were leaking toxic substances into the sewers, lawns, and even basements of homes near the site. The publicity generated by these citizens forced the state to perform formal health studies and in 1978, the state closed the school and relocated some of the residents living closest to the canal. Over 700 remaining residents finally convinced the federal government to declare the entire region a disaster area, resulting in almost all of the families being relocated.

The site has been capped and a drainage system installed to remove the toxic substances as they leak out. The EPA spent about $275 million of taxpayers' money to clean up the site. In 1990 the EPA declared part of the site suitable for inhabitation and the area has been renamed Black Creek Village. The **Office of Technology Assessment** predicts that a "cleaned-up" hazardous waste site will probably once again become unsafe.

LUST This is an acronym for "leaking underground storage tanks," which are a common source of **groundwater pollution.** It is not known how many of the 5 million underground tanks containing **hazardous wastes** and fuels are leaking, but about 5 percent of those tanks containing gasoline are believed to leak, contaminating the groundwater.

macronutrients See *Nutrients, essential.*

Magnuson Fishery Conservation Management Act This act sets U.S. fishing quotas. The quotas are set by government officials and members of the fishing industry. Many environmentalists believe these decisions should be made by marine biologists, who understand the scientific facts about fish populations, and an environmentally literate public.

Malthus, Thomas Malthus (1766–1834) was an English clergyman and economist, famous for his theory about population growth. His theory was put forth in the "Essay on the Principle of Population," which stated that the human population will increase faster than food supplies. It continued, that unless fertility is controlled by late marriage or celibacy, disease and famine would control the growth of populations. Neo-Malthusians believe that population growth should be controlled by birth-control measures. (See *J-curve, S-curve, Carrying capacity*)

Maraniss, Linda Linda Maraniss organized the "Texas Coastal Cleanup," which annually removes 250 tons of garbage from almost 200 miles of Texas beaches. This program and many others around the country not only pick up the trash but record what they find, most of which is plastics. These records helped persuade Congress to ratify an international treaty prohibiting ships from dumping plastic into the sea. By law, these vessels must bring the plastic back to the shore.

  Through these types of programs, the beaches are cleaner, and marine organisms such as whales, birds, and sea turtles are safer. Coastal and shoreline cleanups have become annual events in numerous regions around the country and the world. (See *Plastic pollution, Cruise-ship pollution, Center for Marine Conservation*)

marine ecosystems Aquatic ecosystems are divided into **freshwater** and marine ecosystems. Marine ecosystems have a high concentration of dissolved salts (salinity). Ninety-seven percent of all the surface water on our planet is in the oceans. The concentration of salts in the open ocean is about 35 parts per thousand (ppt), but varies greatly elsewhere. For example, the Red Sea, which has no source of fresh water, reaches 46.5 ppt while the Baltic Sea contains 12 ppt, due to the large inflow of fresh water. The study of marine environments is called oceanography.

Most marine ecosystems are most easily described if divided into two groups, depending upon their distance from the continental shores. The region within a few miles of shore, above the continental shelf, is called the ***neritic zone*** (also called the coastal zone), while the water beyond the continental shelf is the ***oceanic zone.***

Most of the oceanic zone is relatively nonproductive because few nutrients ever make it out to open sea. The most productive marine ecosystems are found in the neritic zone, which includes many specialized habitats. (See *Pelagic marine ecosystems, benthic organisms*)

Marine Mammal Protection Act Established in 1972, this act protects marine mammals, including dolphins, whales, and sea otters, among others in U.S. waters. It contains regulations regarding the use of ***purse seine nets, driftnets*** and ***gillnets.*** It is up for reauthorization in 1993. (See *GATT, Turtle excluder device*)

marsh A wet, unwooded area with dense vegetation, mainly grasses. (See *Wetlands*)

mass transit One of ***urbanization***'s biggest dilemmas is that of transporting vast numbers of people to and from work and around the city. The automobile in many countries chokes its major cities to a halt twice each day and causes ***air pollution.*** Mass transit is an alternative to the automobile. It includes ***light-rail*** and buses. These forms of transportation get people where they must go but conserve energy, save land (that would be used for roads), and reduce air pollution. Unfortunately, mass-transit construction and upkeep costs are very high and often difficult to fund. (See *Bicycles, Intelligent vehicle–highway system*)

meadow A moist, level area with a continuous growth of grasses and other nonwoody (herbaceous) plants.

megalopolis A large, continuous, densely populated region, where ***urban sprawl*** has connected many cities, forming one large city complex, is called a megalopolis. Some examples are the

Boston to Washington DC corridor, the eastern seaboard of Florida, and the southern coastline of California. Also, the region from London to Dover in England and the Toronto-Mississauga region of Canada are considered megalopolises. (See *Urbanization, Ecocities, Greenways, Urban open space, Mass transit*)

megawatt (MW)   Equal to one thousand kilowatts (one million ***watts***).

meltdown   The coolant (usually water) in a ***nuclear reactor*** is responsible for keeping the radioactive fuel rods and the entire system from overheating. If the coolant isn't working properly due to leaks or other malfunctions, the fuel rods and surrounding structures overheat and physically melt, releasing ***radioactivity.*** Meltdowns are supposed to be prevented automatically by sophisticated monitoring equipment, by plant operators, or by both. (See *Nuclear power, Three Mile Island, Chernobyl*)

mercury poisoning   Mercury poisoning can occur by the direct intake of mercury, but usually occurs when contaminated organisms, such as fish that have high concentrations of mercury in their bodies, are eaten and accumulate in the body over time. Mercury poisoning can result in acute illness and disabilities such as numbness and garbled speech, or mental retardation. (See *Heavy-metal pollution, Bioaccumulation, Biological amplification*)

metabolism   Refers to all the chemical processes that occur in a living organism. (See *Toxic pollution*)

metamorphosis   The process of obvious change in form during the development of an organism; for example, a caterpillar building a cocoon and emerging as a butterfly.

meteorology   The study of the ***atmosphere,*** especially as it pertains to weather.

methane digester   Refers to a device that converts ***biomass*** (animal wastes and plants) into methane gas to be used as fuel. The process uses ***anaerobic organisms*** (bacteria), which feed on biomass and produce methane gas. The gas is separated, stored, and then used for heating and cooking. There are millions of these mini–power plants used in China and India. (See *Biofuel, Biomass energy*)

methane gas Methane gas is created naturally as a waste product of anaerobic bacteria (living with little or no oxygen). These bacteria produce methane gas in waterlogged soils and wetlands, but also in human-produced environments such as rice paddies and landfills. The digestive system of animals such as cattle and sheep (ruminant animals) also contain these bacteria and produce methane gas. A single cow belches out 100 gallons of methane gas a day. The microbes in the guts of termites, which digest wood, also produce methane.

Methane is produced for fuel in some parts of the world and burned in **methane digesters.** Methane gas is also a **greenhouse gas** and contributes to **global warming.** About 12 percent of global warming is attributed to increases of methane in the atmosphere.

The top ten sources of methane in our atmosphere are as follows: **wetlands** (20.2%), rice fields (19.4%), cud-chewing animals (14%), biomass fires, such as burning forests (9.7%), oil and **natural gas** pipeline leaks (7.9%), termites (7%), coal mining (6.2%), **landfills** (6.2%), animal wastes (5%), **sewage** (4.4%).

micronutrients See *Nutrients, essential.*

migration Refers to the movement of populations from one area to another. When speaking about human populations crossing international borders, it is called **emigration** and **immigration.** Animal populations often migrate annually as part of their normal behavior patterns. Some migrate for breeding purposes, others simply to avoid harsh conditions. Migratory birds use established **flyways** to cross continents.

mimicry Refers to one organism resembling another organism, usually as a method of protection from predators. (See *Aggressive mimicry, Batesian mimicry, Muellerian mimicry, Predator-prey relationship*)

mineral exploitation Minerals such as iron, nickel, and copper ore are a necessity, but mining these and other resources often impacts the environment. **Recycling** manufactured products reduces the demand for mineral exploitation and helps minimize the environmental damage. To better understand why recycling is important, it is useful to understand how mineral exploitation harms the environment.

Exploration for new mineral deposits often occurs in national parks and other nature reserves, which damages these natural settings and eliminates access by the public. When mineral deposits are located, mining for the ore destroys the area; this is usually due to **strip mining** operations or **subsurface mining.** Crushing the rock to extract the minerals produces large quantities of "leftover" rock, called mining tailings, which are unsightly, result in erosion, and may release harmful substances. Some mining operations use water during processing that becomes contaminated before it is released back into the environment. Many manufacturing processes release contaminants into the air and waste products into the water and soil.

Recycling materials such as aluminum, glass, and many others, and **source reduction**—to minimize the quantity of minerals—minimize the harm produced by mineral exploitation. (See *Natural resources*)

Mineral Policy Center   *Mineral exploitation,* most of which is poorly regulated, causes environmental damage. The Mineral Policy Center is an organization dedicated to cleaning up the impact of mining in America and assuring that the public has a say in the process. Its members are concerned with the destruction of land for mining and the **toxic wastes** that are often a result of these mining operations. Its board of directors is filled with members of many well-known **environmental organizations** who lend their support to the Center's goals. A major goal of the Center is to replace the **Mining Law of 1872.** Write to 1325 Massachusetts Avenue NW, #550, Washington DC 20005.

mini-mill industry   See *Automobile recycling.*

Mining Law of 1872   The General Mining Law was signed into law by President Grant in 1872 and is still in effect, basically in its original form, even today. The fact that it has not been revised for over 120 years is testimony to how the federal government can be pressured into protecting the needs of special interest groups, in this case the mining lobby, instead of protecting the American public. This law allows any person who discovers a valuable mineral on public lands to lay claim to that land. The claim holder pays no rent to the federal government, but can rent or lease the land to others and charge them rent or royalties. Furthermore, the claim holder can then purchase the land and its contents from the government for five dollars per acre! Over 1.2 million of these claims have been filed for over 25 million acres of what was public land; 2,000 of these acres are in national parks.

Large tracts of these claimed lands never get mined, but instead are developed and built upon at huge profits for the owners. Efforts to amend or rewrite this law have failed for years. The **Mineral Policy Center,** along with some members of Congress, has been fighting to change the law. (See *Mineral exploitation*)

MNS Online MNS Online is the official **ecolinking** electronic (computer) bulletin board. Environmental reports, programs, and news can be downloaded. Online forums are available on many environmental topics. For information, contact Ecolinking at P.O. Box 463, Schenectady, NY 12301-0463. (See *Computer networks, environmental; Computer software, environmental; News Service, Environment*)

money market funds, environmental There are many money market funds that specialize in investing monies only in institutions they deem socially responsible when it comes to the environment. Companies with histories of environmental violations, for example, are excluded from their portfolio. A few of the many funds available include: Working Assets Money Fund, (800) 223-7010; Calvert Social Investment Fund, (800) 368-2750; John Hancock Freedom Environmental Fund, (800) 225-5291; Global Environment Fund, (202) 466-0529; Parnassus Fund, (800) 999-3505; and PAX World Fund, (800) 767-1729. (See *Payroll deduction, environmental; Ethics, environmental; Environmentalist*)

monoculture When a single species is planted over a large area of land, it is called a monoculture. Farmland covered with thousands of acres of corn or wheat are examples of monocultures. Monocultures are ultra-simplified ecosystems. A single species of plant exists where a grassland or forest ecosystem with hundreds or thousands of species once existed. These manmade ecosystems are a very efficient method of growing crops, but are highly vulnerable to change and must continually be protected and provided for by manmade means. Monocultures need extraordinary amounts of **pesticides** to protect and **fertilizers** to provide.

For example, unwanted weeds naturally try to invade monocultures, so **herbicides** are used. Insect pests that feed on the crop have a field day in a monoculture since its food source is so abundant, so **insecticides** are used.

Insects reproduce rapidly and develop resistance to these insecticides through the process of **natural selection,** so greater dosages of insecticides must be used just to maintain the same level of control. Monocultures also invite plant diseases to establish themselves, so **fungicides** are used to defend the crop.

Since crops grown in monocultures are constantly harvested, there is little if any part of the plants that remain to nourish the soil for the next crop, so the soil must constantly be enriched with **fertilizers.** Many **organic farming** practices reduce the need for pesticides and fertilizers. (See *Sustainable agriculture, Integrated pest management, Irrigation*)

Montreal Protocol, The   Originally signed in 1987 by 24 countries including the U.S., Canada, Japan, and the European nations, and later amended, this document pledges that the signatories will phase out the use of all **CFCs** by 1999. (See *Ozone-depleting gases*)

moon power   Twice each day, the gravitational pull from the moon causes the ocean waters to flow in and out along the coasts, resulting in the tides. If the tides flow through narrow inlets, the water can be channeled through turbines, which are turned by the moving water and generate electricity. Two of these moon power plants exist—one in France and the other in Canada. Although they are an excellent source of clean energy, there are believed to be few sites suitable for the construction of these moon power plants to be a major player in supplying the world's energy needs. (See *Alternative energy, Ocean thermal energy conversion, Solar power, Solar ponds, Hydroelectric power*)

morbidity   Morbidity refers to disease within a **population** and its effect on the population. Information about the frequency and distribution of a disease can help control its spread and determine its cause. (See *Mortality*)

more developed countries (MDC)   Defined by the United Nations as a country with high industrialization and a high gross national product (compared with **less developed countries**). There are 33 MDCs, including the United States, Canada, Japan, Australia, New Zealand, and all the countries in western Europe. Most of these countries have favorable climate and fertile soils.

About 23 percent of the world population lives in MDCs, but about 80 percent of the world's energy resources are used by them. The annual population growth rate of these countries is about .5 percent, which is considered slow (compared with 2.1 percent for the LDCs, which is considered very rapid). The average GNP for those living in an MDC is about $15,800 per year compared with someone in a LDC who averages about $700 per year. (See *Standard of living*)

mortality Mortality refers to the probability of dying within a population. Everyone dies sooner or later, but the probability of dying is related to factors such as age, sex, race, occupation, social class, and many other factors. The incidence of death usually reveals a great deal about a population's **standard of living.**

Mortality is usually measured as the "crude death rate." This is calculated by taking the number of deaths at the midpoint of a specific period of time and dividing it by the population at the beginning of that period, and then multiplying by 1,000. The resulting figure is expressed as deaths per 1,000.

For example, if there were ten deaths in a population of 3,000 individuals in one year, the death rate for that population is 3 per thousand, per year (10 divided by 3,000, multiplied by 1,000). Mortality figures are generated for all types of organisms, including humans. For example, the mortality (crude death rate) in the U.S. in 1990 was 9 per thousand. (See *Population dynamics*)

motor oil recycling Do-it-yourself automobile motor oil changes produce over 200 million gallons of used oil each year in the U.S., with most of it illegally dumped and not recycled. One gallon of drained motor oil can be recycled to produce the same amount of fresh clean oil as 42 gallons of crude oil out of the well. Most towns and cities have their own motor oil collection sites, facilities, or regulations. (See *Recycling, Automobile recycling, Tire recycling*)

mountain Refers to a large land mass at least 300 meters (slightly less than 1,000 feet) above the surrounding terrain with a summit narrower than the base. Smaller structures are usually considered hills.

muck Refers to a fine grain soil, saturated with water and of a thick consistency. It is dark in color, containing a high concentration of well-decomposed *organic* matter such as plant and animal wastes. Contains less organic matter than *peat*. (See *Sludge, Ooze*)

Muellerian Mimicry   A form of *mimicry* in which many species of distasteful or poisonous organisms all look similar, making it simple for *predators* to avoid. For example, there are many species of poisonous butterflies all with similar coloration to warn birds to stay away. (See *Directive coloration, Aposematic coloration*)

Muir, John   John Muir is considered by many to be America's greatest naturalist and conservationist. He was largely responsible for the establishment of many national parks including Sequoia, Yosemite, Rainier, and the *Grand Canyon*. After attending the University of Wisconsin, he worked on mechanical inventions, but abandoned that career to devote himself to nature. He walked from the Midwest to the Gulf of Mexico and kept a journal that was published in 1916. In 1868, he went to the Yosemite Valley and took numerous trips to Nevada, Utah, Oregon, Washington, and Alaska. Through his travels, he theorized that the formations in Yosemite were due to glacial erosions, now accepted as the cause. In 1876, he urged the federal government to adopt a forest conservation policy. In 1892, Muir helped found the *Sierra Club,* which continues to create and preserve our national parks. (See *Ethics, environmental; Philosophers, environmental; Ice ages*)

municipal solid waste (MSW)   Simply put, MSW is *garbage.* It specifically refers to a municipality's solid waste, which is usually collected by the town or private company and delivered to, or dropped off at, a collection site. In the U.S., about 160 million tons of solid waste are produced each year, or about 3 pounds of garbage per person per day and increasing all the time.

The largest component of U.S. solid waste is paper products, which makes up about 40 percent. Yard waste (grass clippings and the like) comprise about 17 percent; rubber, textiles, and wood products are about 12 percent; and then metal, glass, plastic, and food waste amount to about 8 percent each. *Municipal solid waste disposal* is one of the biggest environmental challenges we face.

Local governments historically were responsible for MSW management, but the federal government became involved in 1965 with the Solid Waste Disposal Act. More recent legislation includes the

1976 Resource Conservation and Recovery Act (RCRA) and amendments to it that provide a program to eliminate open dumping, promote solid waste management programs, establish standards for *landfills* and air emissions, provide grant money for rural communities, and regulate *hazardous waste.* (See *Waste stream, Waste-to-energy power plants, Waste minimization, Source reduction, Recycling, Grass cycling, Plastic recycling*)

municipal solid waste disposal **Municipal solid waste** can be disposed of today in three ways: 1) dumping it in landfills; 2) burning it in incinerators; or 3) recycling it. *Landfills* are the most common, since they are economical and—until recently—thought to be totally safe. About 80 percent of the MSW in the U.S. goes into landfills. Nine percent is incinerated, but new technologies may make *incineration* more popular in the near future. Much emphasis has been placed on *recycling* the waste before it ever gets to landfills or incinerators. At present, however, only 11 percent is recycled. Japan recycles 45 percent of its solid waste, and many other countries have been recycling for many years.

Finally, *source reduction* is designed to reduce the amount of garbage that must be disposed of by simply producing less to begin with. (See *Waste stream, Waste-to-energy power plants, Waste minimization, Source reduction, Grass cycling, Plastic recycling, Appliance recycling*)

mutagenic Refers to substances that cause genetic defects. (See *Toxic pollution*)

mutualism Mutualism is a *symbiotic relationship* in which both organisms involved benefit. Insect-plant relationships of this type are common. For example, the yucca plant is pollinated by the yucca moth, enabling the continued survival of the yucca plant. The moth lays its eggs in the plant and the young depend on the plant's seeds for nourishment. Both plant and insect survive because of this mutual relationship. *Nitrogen fixation* is the result of another mutualistic relationship between bacteria and legumous plants.

natality Natality is the birthrate of a population. More specifically, it refers to the "crude birthrate," which is calculated by taking the number of births in a specific period of time, such as one calendar year, dividing it by the estimated population at the midpoint of that period, and multiplying by 1,000. The resulting figure is expressed as births per 1,000 for a certain period of time.

For example, if a population of 2,000 individuals has ten births in one year, the natality for that population is 5 per thousand per year (10 divided by 2,000 multiplied by 1,000). Natality figures can be generated for all types of organisms, including humans. For example, the natality (crude birthrate) in 1990 in the U.S. was 16 per 1,000. (See *Mortality, Population dynamics, Population size, Carrying capacity*)

National Audubon Society Founded in 1905, the Audubon Society works to protect wildlife and its habitat primarily through education, research, and political action. The society has over 500,000 members in 500 chapters nationwide, which are involved in numerous local conservation issues. Its *Audubon* magazine is world renowned for its nature photography. The annual membership fee is 30 dollars. Write to 700 Broadway, New York, NY 10003-9501.

National Energy Bill This bill was passed in 1992 by Congress and is designed to make the U.S. electric and **natural gas** system more efficient and provides some incentives for **renewable energy** research and use. The bill, however, provides subsidies to the oil industry and no incentives to reduce our dependency on oil. It also places an increased role on the use of **nuclear power** in the future and reduces many of the safeguards that existed in the licensing procedures of these plants. (See *Energy use*)

National Marine Sanctuaries Act This act protects unique marine ecosystems, such as the **coral reefs** off the Florida coast, from pollution and development. Ten such sanctuaries have been protected by this act, which is about 1 percent of the U.S. coast. (See *Estuary and coastal wetlands destruction*)

National Park and Wilderness Preservation System The National Wilderness Preservation System was created in 1964 when President Johnson signed the Wilderness Act. It established about 9 million acres of federally protected lands. Today there are over 90 million acres of protected wilderness in 546 areas with the vast majority found in Alaska (56.5 million acres). These areas are designated as wilderness, contain no roadways, and are protected to remain as such by law. Wilderness land includes national forests (33.6 million acres) managed by the U.S. ***Forest Service,*** national parks (39.1 million acres) managed by the U.S. Park Service, wildlife refuges (20.7 million acres) managed primarily by the U.S. Fish and Wildlife Service, and other lands managed by the Bureau of Land Management (1.6 million acres). Wilderness areas make up only a small portion of the 634 million acres of all public lands. (See *Grand Canyon, Tongass National Forest, Arctic National Wildlife Refuge, Land acquisition, wilderness*)

National Wildflower Research Center The Wildlife Research Center is the only institution in the nation dedicated to conserving and promoting the use of native plants in North America. This research center studies flowers not only for their beauty but also for the economic and environmental benefits they provide. They have a clearinghouse and a reference library, which contain materials on native plants, propagation and landscaping, and wildflower identification. The organization was created by Lady Bird Johnson in 1982. Write to 2600 FM 973 North, Austin, TX 78725-4201.

National Wildlife Federation (NWF) The NWF, with 5.8 million members and founded in 1936, is primarily a conservation education organization promoting the appropriate use of our natural resources. The group espouses that the welfare of wildlife and that of humans are inseparable and that wildlife is an indicator of the environmental quality of the planet. They work to influence conservation policies at all levels through legislative and legal means. The annual membership fee is 16 dollars. Write to 1400 16th Street NW, Washington DC 20036. (See *Environmentalist*)

natural gas Natural gas, a ***fossil fuel,*** supplies about 17 percent of the world's energy and about 24 percent of the U.S.'s energy. It can be found by itself or often together with crude ***oil*** deposits.

Natural gas is believed to have been formed under conditions similar to those of *oil formation,* but continued to change into a lighter hydrocarbon, methane gas.

About 40 percent of the world's remaining natural gas reserves are found in what was the Soviet Union. The U.S. has only about 6 percent of the world reserves. Almost all of the U.S. demand for natural gas is for home heating.

Natural gas supplies are expected to outlast oil. Known reserves and anticipated resources are expected to last about 60 years at current consumption levels. If new technologies are developed that could extract currently unavailable gas deposits, the supply could last roughly 200 years.

Natural gas is cleaner than other fossil fuels and has a higher heat content. It produces fewer pollutants and less *carbon dioxide* (a *greenhouse gas*) than other fossil fuels. The only major concern is the danger of shipping *liquified natural gas,* since it is highly flammable. Coal-burning plants that generate electricity could convert to cost-effective natural gas–burning turbines.

Many people believe that as oil is phased out, natural gas will fill the void until *solar power* or other *renewable energy sources* are fully developed and implemented. (See *Natural gas extraction*)

natural gas extraction   Similar to crude oil, *natural gas* is brought to the surface by wells. Primary recovery involves tapping and removing the gases by natural forces. Secondary recovery involves pumping air or water into the well to force the remaining gas from the deposit.

Natural gas is composed primarily of methane gas, but also contains small amounts of butane and propane. The methane is separated, dried of moisture, cleaned of impurities, and pumped into pipelines for distribution. It can also be converted into liquified natural gas (LNG) at very low temperatures so it can be shipped by tanker for export to other countries.

The butane and propane are usually separated and liquified to form liquified petroleum gas (LPG), which is stored in pressurized tanks and shipped to rural areas that don't have natural gas pipelines. (See *Strip mining, Subsurface mining, Oil recovery*)

natural history   Refers to the study of nature, including all natural objects, living and nonliving, and natural processes. (See *Environment*)

natural insecticides Although almost all the **insecticides** applied each year are synthetic insecticides, there are a few found in nature that need not be created in the lab. They are divided into botanicals, which are derived from plants, and mineral pesticides that come directly from the earth. Examples of botanicals include pyrethrin, which is extracted from chrysanthemums. Pyrethrin was the original model for many synthetic pyrethroid pesticides. Nicotine, from the tobacco plant, and rotenone, from legumes, are also botanicals.

Mineral pesticides include boric acid, which contains boron. Sulfur and copper have been used as natural **fungicides.** Other natural methods include the use of smoke to repel insects, and light oil sprayed on crops. (See *Biological control, Biological pesticides, Integrated pest management*)

natural resources Natural resources refer to substances, structures, or processes that are often used by people but cannot be created by them. For example, the sun, land, and the oceans are natural resources, and their uses are obvious. Iron ore is a natural resource since we use it to make steel, and the **Grand Canyon** is a natural resource since it is a natural wonder and a popular tourist attraction.

Natural resources can be either renewable or nonrenewable. Renewable resources include the sun, soil, plants, and animal life, since they all naturally perpetuate themselves. Some of these renewable resources, such as the sun, are used as **renewable energy** sources. Nonrenewable resources are those which do not perpetuate themselves. If continually used by humans, they will someday be "used up." For example, the supply of a mineral such as iron ore is finite and will become exhausted someday. Most of the world's energy needs come from fossil fuels, which are **nonrenewable energy** sources and will be gone in the future.

Using natural resources often takes an environmental toll on our planet by causing some form of pollution or damage. Extracting coal destroys the land, and burning it for electricity pours tons of pollutants into the air. Using our natural resources should be balanced between our needs and the impact on our environment. (See *Mineral exploitation*)

Natural Resources Defense Council (NRDC)
The NRDC is dedicated to protecting our air, water, and food supplies. Through legislation and research, it has led the fight against **acid rain** and fought for proper enforcement of the **Clean**

Air Act. Through its campaign Mothers and Others for Pesticide Limits, it helped eliminate the use of *alar* on apples. The NRDC also began forest management reform and fought federal coal-leasing and oil-drilling policies that threatened wildlife. Write to 40 West 20th Street, New York, NY 10011.

natural selection   Refers to the process, originally proposed by Charles Darwin in his classic book *Origin of Species,* in which a species gradually adapts to its environment. Natural selection occurs when individuals of a species with "genes" better suited to their environment survive, and those less suited, die. As generations pass, those with the best genetic makeup are naturally selected for survival, carrying these genes to future generations. Over long periods of time, entirely new species may evolve, perfectly suited for their environment.

It is difficult to comprehend processes that normally take thousands or millions of years to occur, but there are short-term examples of natural selection, such as the classic study on peppered moths. Peppered moths rest during the day on tree trunks and other surfaces. If a bird sees the moth, it will most likely be eaten. Prior to the mid 1900s, peppered moths were colored to blend in nicely with the light-colored tree trunks. This is called cryptic coloration.

In industrial regions of England in the mid 1900s, the tree trunks became blackened due to the pollution caused by coal-burning power plants. The light-colored moths began to stand out on the trees and were easily made into a meal by birds. Those moths with darker coloration began to be naturally selected to survive since they were less visible. In these polluted areas the dark variety of the species became the norm and the light variety a rarity. (See *Speciation*)

nature centers   Many urban centers have so little of the natural environment left, that people, especially children, have little or no knowledge about the wilds of nature. Perhaps for this reason, nature centers have become popular near many big cities. A cross between a park and a *zoo,* these educational facilities teach people about natural environments. Nature centers are often funded by state or local governments or nonprofit organizations. (See *Urban open space, Touchpools, Environmental education*)

Nature Conservancy, The (TNC)   The Nature Conservancy is the leader in the purest form of environmentalism— preserving and protecting unspoiled lands. The Nature Conservancy, established in 1951, identifies, acquires, and manages unique natural environments so they cannot be destroyed by human intervention. It has helped preserve over 5 million acres in 50 states, Canada, and other countries. This organization has purchased and now manages over 1,000 sanctuaries, most of which are home to endangered species. TNC has a nationwide network of local and state chapters. The annual membership fee is 15 dollars. Write to 1815 North Lynn Street, Arlington, VA 22209. (See *Land acquisition, wilderness*)

negawatts Refers to "negative watts," meaning electricity that has been saved as opposed to used or wasted. Coined at the Rocky Mountain Institute, the word helps define and encourage saving electricity as well as simply using it. For example, how much money a company can save by finding negawatts in its buildings and factories. (See *Energy use, Watt*)

nematodes Nematodes, also called roundworms, are microscopic, unsegmented worms found in large numbers in almost all habitats. A few are **parasitic** on humans.

neritic zone (coastal zone) **Marine ecosystems** can be divided into those found in the open ocean (**oceanic zone**) and those found along the coast, called the neritic zone. The neritic zone includes all the waters above the continental shelf, beginning at the shore and extending out to sea several hundred miles in some areas. The immediate coastal area is continually affected by the tides and contains some of the most productive environments on earth.

These coastal waters are home to many fish, clams, oysters, crabs, sponges, anemones, jellyfish, among others. In much of this water, the sunlight penetrates to the bottom, allowing plants to attach to the substrate, providing shelter for many other organisms. There is a great deal of variety in the types of ecosystems in these areas, since there are many types of shorelines. Rocky shores have very different ecosystems compared with sandy shores.

The coastal zone has many unique, highly productive habitats, including **estuaries, coastal wetlands,** and **coral reefs.** Some areas in this zone exhibit **upwelling,** which provides a habitat for interesting ecosystems. (See *Littoral zone. Aquatic ecosystems*)

net primary production (NPP) NPP is used to compare the plant productivity of different ecosystems. NPP is the rate at which plants build organic molecules containing useable energy (*photosynthesis*), minus the amount of energy used up by the plants themselves in *respiration* in a given area at a given time. It is often measured in dry grams per square meter, per year (g/m²/yr). *Wetlands* and *tropical rain forests* have the highest NPP levels, ranging from 1,000 to 3,500 g/m²/yr, while the lowest are found in *tundra* (with 10 to 400) and *deserts,* which range from 0 to 250 g/m²/yr. (See *Energy pyramid*)

News Service, Environment (ENS)   The Environment News Service is an international news agency dedicated to gathering and disseminating worldwide environmental news. This news is delivered by electronic networks and fax to online (computer) services, magazines, newspapers, and organizations. This service routinely taps a resource pool of thousands of experts in environmental fields to help find, decipher, and explain the news. For information, call the Environment News Service at (604) 732-4000 in Canada. (See *MNS Online, Econet*)

niche   Every *species* is unique from every other species in some way. They each place their own demands on an *environment* and contribute in their own way to that environment. This specific role in the environment is called an organism's ecological niche. A niche is different from the organism's *habitat,* since it is more concerned with the impact of the organism on the environment than with the environment's impact on the organism, as with habitat. (See *Keystone species*)

night soil   Refers to human manure sometimes used as an *organic fertilizer* in less developed countries.

NIMBY syndrome   NIMBY is an acronym for Not In My Backyard. It refers to local opposition to new *landfills, incineration* plants, and *hazardous waste disposal sites,* among others. When people speak about running out of room for waste, such as landfills, they really mean running out of places where people will allow them to be built. Now that people realize the inherent dangers that come with many

of these facilities, they take action to prevent them. (See *NOPE, NIMEY syndrome*)

NIMEY syndrome NIMEY is an acronym for Not In My Election Year, which is the politicians' counterpart to the electorate's **NIMBY syndrome.**

NIOC syndrome NIOC is an acronym for Not In Our Country. Some less developed countries accept **hazardous wastes** and **toxic wastes** from other countries as a way to generate income. As these countries become more aware of the health risks involved in accepting this waste, some now profess the NIOC syndrome, and refuse to act as a hazardous waste dumping ground. (See *NIMBY syndrome, NIMEY syndrome, NOPE, YIMBY—FAP*)

nitrogen compounds Nitrogen compounds are one of the five primary pollutants that contribute to **air pollution.** The most common of these compounds are nitrogen oxide and nitrogen dioxide. As with other primary pollutants, nitrogen compounds come primarily from automobiles and electric power–generating plants. These compounds play a major role in the production of **secondary air pollutants** that create **photochemical smog.** They also contribute to the development of **acid rain.**

nitrogen cycle Nitrogen is important to all organisms. It is a primary component of proteins. Even though 78 percent of the air is composed of nitrogen, green plants are incapable of using it in this form. Most nitrogen becomes available to green plants (and therefore to all life) thanks to the process of **nitrogen fixation,** in which cetain organisms take nitrogen from the air and convert it into biologically useful nitrogen that green plants can absorb through their roots and incorporate into their tissues. This nitrogen is then passed through **food chains** to animals.

When plants and animals die, the nitrogen found in their bodies is also made available to green plants, but it must first be broken down into a useable form. This is done by a specific type of bacteria, which breaks down complex biological molecules that contain nitrogen into a form that can be absorbed by the roots of green plants.

Two other interesting aspects of the nitrogen cycle include other kinds of bacteria (denitrifying bacteria) that convert organic nitrogen in the soil directly into free nitrogen found in the air to begin the cycle over again. Finally, lightning also plays a minor role by combining nitrogen in the air with oxygen to form a useable form of nitrogen, which falls to the ground with precipitation and becomes available to plants. (See *Biogeochemical cycles*)

nitrogen fixation   Seventy-eight percent of the air we breathe is composed of nitrogen gas. Nitrogen in the air, however, cannot be used by organisms, so it must change form (be "fixed") before becoming accessible to plants, which occurs during the process of nitrogen fixation. This process occurs in part due to a ***symbiotic relationship*** that exists between ***bacteria*** that live in the roots of plants such as bean, pea, and clover (legumes). The bacteria produce swellings or nodules in the plant's root. The bacteria in the nodule take nitrogen from the soil (actually air that permeates the soil) and incorporate it into their own cells. When the bacteria in the nodules die, the organic nitrogen becomes available to the plants. When eaten, the nitrogen is passed along the ***food chain***.

There are other nitrogen-fixing bacteria that live freely in the soil and water. Some trees and grasses have a symbiotic relationship with nitrogen-fixing ***fungi***. (See *Nitrogen cycle*)

nitrous oxide   Nitrous oxide gas is produced from the breakdown of fertilizers and livestock wastes and from burning fossil fuels and other forms of ***biomass***, such as wood. Nitrous oxide is a ***greenhouse gas*** and contributes to ***global warming***. About 6 percent of global warming is attributed to increased concentrations of this gas in the atmosphere. (See *Air pollution*)

no net loss   This phrase is used by local, state, or federal agencies regarding the loss of environmentally sensitive lands. It means the following: If human activities destroy a natural resource such as a ***wetland***, humans will "create" an artificial replacement for it elsewhere, such as a manmade wetland. This policy is viewed with great skepticism by environmentalists, among others. (See *Greenspeak, Greenscam, Estuary and coastal wetlands destruction*)

noise pollution   Noise pollution refers to any noise that annoys or harms individuals. At certain levels, noise can pose a health risk. Noise pollution is measured in decibels (db). To take into account the pitch of the sound, dbA is used, which indicates the A scale.

The following examples indicate when noise becomes a problem: a whisper (20 dbA), normal conversation (50 dbA), a vacuum cleaner (70 dbA); a garbage disposal or nearby train can cause hearing damage (80 dbA); a diesel truck or food blender for extended periods can cause permanent hearing damage (90 dbA); a chain saw or thunderclap causes pain (120 dbA); and a jet taking off less than 100 feet away can rupture an eardrum (150 dbA).

In addition to immediate ear damage, headaches and many other ailments have been blamed on noise pollution. The Noise Control Act of 1972 was the first federal attempt to control noise pollution in the U.S. Local and state governments have enacted their own laws governing noise pollution. (See *Aesthetic pollution*)

nongovernmental organization (NGO) NGOs include any organization that is neither governmental nor for-profit. They include a vast array of organizations, including numerous environmental advocacy groups. Some are small, grassroots groups working at the community level; others are regional, national, or even international. Most are membership oriented and stress volunteerism. The United Nations has worked closely with many NGOs to help establish communications on a global scale. (See *Environmental organizations*)

nonpoint pollution See *Pointless pollution.*

nonrenewable energy Industrialized nations depend primarily on **fossil fuels,** which include coal, oil, and natural gas, as their source for energy. These fuels are considered nonrenewable since they are no longer being replenished and will become depleted over time. **Renewable energy** sources, however, are considered to be available for an infinite length of time, since the supply is continually replenished. (See *Oil formation, Coal formation*)

nontarget organism **Pesticides** are designed to kill a specific organism—the pest. Any organism affected by a pesticide that is not the targeted pest is called a nontarget organism. Since most pesticides kill more nontargeted organisms than the pests, some people prefer to use the term "biocide" instead of pesticide, since this refers to killing many forms of life.

noosphere Pertains to that portion of the **biosphere** that is affected by human influence. (See *Anthropogenic stress*)

NOPE   Acronym for Not On Planet Earth, which is the logical extension of the **NIMBY syndrome.**

North American Association for Environmental Education (NAEE)   Founded in 1970, the NAEE specializes in improving the environment through education and has been a major force in training environmental educators at all levels, kindergarten through university level. The 1,000 + members include teachers, writers, naturalists, managers, film makers, and just about anyone involved with education and concerned about the environment. The NAEE sponsors an annual week-long conference. The annual membership is 35 dollars. Write to 1255 23rd Street NW, #400, Washington, DC 20037. (See *Environmental education*)

Northern Spotted Owl   The Northern Spotted Owl has become a symbol of how the environment and the economy can be at odds with one another. In June of 1990, the Northern Spotted Owl was placed on the **threatened species** list. This bird lives in the **ancient forests** of the Pacific Northwest and, like most predatory birds, requires large expanses of land to hunt for food. Only 10 percent of these forests remain, most of which are on **Forest Service**–managed lands.

The only way to protect the owl is to protect these forests from logging. Members of the logging industry believe that protecting these lands would cost over 90,000 logging jobs in the region. Opponents believe this number simply represents a continuation of an already ongoing decline in the number of logging jobs. (See *Economy vs. environment, Sustainable agriculture, Endangered Species Act*)

nuclear fission   Nuclear fission is the process of splitting an atom to release energy. This energy can be released for destruction as is the case with nuclear bombs, or it can be controlled and harnessed as an energy source called **nuclear power.**

The nucleus of an atom is composed of neutron and proton particles, which are bound together by energy. In most elements, the nucleus remains stable, meaning the particles and the energy contained within, stay bound in the nucleus infinitely. Some elements, such as uranium and plutonium, are radioactive, meaning the nucleus is unstable and particles are released from the nucleus—a process called decomposition. When this happens, the energy that usually binds the particles together is released along with the particles. In its simplest sense, this is nuclear power. The decomposition of nuclei can be harnessed to produce useful power in *nuclear reactors* to generate electricity.

nuclear fuel cycle A *nuclear reactor* uses radioactive materials (usually uranium), which must be mined, milled, enriched, fabricated, used, and, finally, disposed of. Low-grade uranium ore is first mined and then goes through a milling process, which involves crushing and treating it to concentrate the uranium. Once milled, the mixture is called "yellowcake." The crushed uranium ore that remains is called mining tailings.

Yellowcake must then go through an enrichment process to increase the amount of radioactivity, so it can sustain a chain reaction. Once enriched, it goes through a fabrication process in which it is turned into pellets that are used to fill fuel rods about 4 meters (13 feet) in length. The fuel rods are placed within the reactor core. The energy released by nuclear fission is used to generate electricity in the *nuclear power* plant.

After three to four years, the fuel rods can no longer support the chain reaction and must be removed from the plant and disposed. (See *Radioactive waste disposal*)

nuclear fuel cycle hazards Each step in the *nuclear fuel cycle* has inherent dangers. Uranium miners are exposed to low-level radiation from the ore. Chronic exposure to this radiation is believed to increase the miners' likelihood of getting certain illnesses such as lung cancer. During the milling process the ore is crushed, leaving low-level radioactive *tailings,* which can harm organisms and leach radioactivity into the *groundwater.*

During the enrichment and fabrication processes, individuals must be protected from exposure to the high-level radiation released by the fuel. Once in the reactor, the intense levels of radiation must be contained. Radiation leaks can be catastrophic as seen in the *Chernobyl* disaster.

The biggest problem associated with *nuclear power* is disposing of the used, but still highly radioactive, fuel rods. *Nuclear waste disposal* technology is just beginning to be developed. In addition, every step of this cycle involves transporting the radioactive material from one location to another, which carries the danger of an accident and the release of radioactivity.

nuclear fusion Conventional *nuclear power* plants use *nuclear fission,* the splitting of atoms, to release energy. However, when nuclei are combined (fused) together, energy is released.

The sun is powered by nuclear fusion. The process involves combining two hydrogen nuclei to form helium. Controlling this process has been elusive to scientists, who have been studying it for decades. Nuclear fusion might someday supply the world with all the energy it needs, but this will not be available for decades to come, if ever.

nuclear power Nuclear power is an **alternative energy** source that became popular in the 1970s and proliferated in the early 1980s. In 1989, there were 110 nuclear power plants in the U.S. providing about 20 percent of our energy needs. There were predictions that by 2010, 40 percent of our energy would be generated by nuclear plants.

There are many advantages to nuclear power. It does not cause **air pollution** or release **greenhouse gases** into the atmosphere and causes minimal **water pollution.** But the growth of nuclear power has dramatically declined in the U.S. during the late 1980s due to two major factors: safety and cost. The **Three Mile Island accident** in 1979 raised public awareness of safety problems with nuclear reactors. The **Chernobyl** disaster renewed and re-enforced these fears in 1986. The Nuclear Regulatory Commission has projected a 45 percent chance of a serious nuclear accident in the U.S. within the next 20 years. (This does not necessarily mean radioactivity would escape.)

Besides accidents, there are serious concerns about **nuclear waste disposal** of the spent radioactive materials. Additional safety issues include inherent building flaws and associated containment problems that may allow the release of radioactive gases. Some plants are showing signs of premature aging, which may force them to close before their 40-year licenses expire.

In addition to safety problems, the costs of building nuclear plants and the ongoing cost of producing electricity have been higher than originally thought. Many plants were never completed due to cost overruns. Those in service are generating power at somewhat high prices but are still competitive.

Although the future of nuclear power in the U.S. is debatable, France intends to generate almost all of its electricity from nuclear power during the 1990s. New designs for smaller, safer, and less expensive reactors are being studied. These new technologies will generate only 100 to 300 **MW** of power compared with today's reactors generating 1000 MW. These new designs use passive safety features, meaning they use simple natural forces such as gravity to shut down the reactor in an emergency, instead of human operations and complex electronic computer systems, which might fail.

Even if the safety and cost issues can be corrected with a new reactor design, a definitive solution to the disposal of spent nuclear waste must be found. Safety and costs are the primary reasons why no new plants have been ordered in the U.S. since 1978 and all plants ordered after 1973 have been canceled. If this trend continues, with existing plants aging rapidly and no new plants built, the contribution of nuclear power to U.S.'s future energy needs will decline over the next few decades.

nuclear reactor safety and problems   The accidental release of radiation from a **nuclear reactor** could cause catastrophic results. Nuclear power plants use many safety features to prevent accidents, including: systems that automatically shut down the reaction in case of an emergency such as a loss of coolant; concrete and steel walls surrounding the core of the reactor to prevent the release of radiation; concrete and steel containment walls around the entire reactor to contain radiation if the core ruptures; emergency electrical systems to assure that the cooling system doesn't fail, and emergency cooling systems in case it does fail. Even with all these safety features, plus others, accidents have happened. Accidents at **Chernobyl** and **Three Mile Island,** in which radiation was released, were caused by a series of equipment failures and/or operator errors.

Besides the primary danger of the release of radiation from the reactor, other problems exist. About one-third of the heat generated by nuclear power plants is used to generate electricity, while the other two-thirds is carried away by the coolant, causing **thermal water pollution.** Finally, the plant must go through **decommissioning,** which means it is taken out of service. This is difficult and costly since the facility is contaminated with radioactivity.

nuclear reactors   A nuclear reactor controls **nuclear fission** and uses the energy it releases to generate electricity. Nuclear reactors allow a chain reaction of nuclear fission to occur in which one atom is split and the particles released are used to split other atoms, etc.

A supply of a radioactive element such as uranium-235 is placed in a reactor in the form of fuel rods. A slow-moving neutron from another element strikes the uranium, initiating the chain reaction. The energy released from the chain reaction super heats surrounding water, which is removed from the reactor and used to drive turbines and generate electricity. (The water also acts as a coolant in the system, keeping it from having a meltdown in which the fuel rods melt from overheating.) The speed of the chain reaction is controlled by control rods containing cadmium and boron, which absorb the nuclei.

There are three different types of nuclear reactors: *light-water nuclear reactors, heavy-water nuclear reactors,* and *high temperature, gas-cooled nuclear reactors.*

nuclear waste disposal   A typical nuclear power plant produces 30 metric tons of radioactive waste per year. There are now 14,000 metric tons of uranium waste temporarily stored in various locations, awaiting a permanent disposal site. Since the radioactive material remains dangerous for tens of thousands of years, this material poses a significant threat to all forms of life, indefinitely.

The U.S. has been trying to find a permanent repository for radioactive nuclear waste materials. By the year 2000, there will be about 40,000 metric tons of high-level radioactive wastes, which includes spent fuel rods from *nuclear reactors.* At the present, most of the spent fuel is stored in temporary cooling ponds, awaiting a final destination.

Congress has designated two locations to become permanent repositories for nuclear waste materials. The first is the *Waste Isolation Pilot Plant (WIPP)* near Carlsbad, New Mexico, scheduled to open in a few years and become the first permanent nuclear waste storage site. Public opposition (*NIMBY syndrome*) has been intense and threatens the site's future. The second is in the *Yucca Mountains* of Nevada, which isn't scheduled to open for many years. The nuclear materials would be injected deep within the geological formations that would hopefully contain the radioactivity indefinitely.

The safety of this method is hotly debated, although many scientists feel it would pose little risk. The EPA, the agency responsible for setting safety standards, has told the Department of Energy, which is responsible for building a repository, that the facility must be able to contain the radioactivity as long as it is still dangerous, which is at least 10,000 years. Apparently, Congress, along with the American public, is fearful of the idea. There are concerns about the stability of both sites. An earthquake shook the Yucca Mountain facility in July 1992.

Other countries are also finding public opposition to building permanent nuclear waste disposal sites and some, such as Sweden and France, have chosen to use temporary sites and study permanent site possibilities at least until the year 2003. (See *Hazardous waste disposal, Hanford Nuclear Reservation*)

nutrients, essential   Almost all of the substances required for life are found in nature as minerals. These minerals enter the living world when they are absorbed by plants, which pass them along ***food chains.*** Organisms need about 40 different substances to survive, called essential nutrients. They can be divided into three categories.

Those required in large quantities every day are the "bulk elements" and include hydrogen, carbon, and nitrogen. The next group is needed to a lesser degree and are collectively called the macronutrients, which include phosphorus, potassium, calcium, magnesium, and sodium. Finally, others are needed in very small amounts and are called the micronutrients or trace elements. These include iron, copper, zinc, and iodine. (See *Biogeochemical cycles, Phosphorus cycle, Carbon cycle*)

Ocean pollution   Oceans have been a favorite dumping ground for human activity. Their vastness results in quick dispersal and dilution. A sad old saying states, "The solution to pollution is dilution." This works deceivingly well until the volume of pollution overwhelms the volume of the water, which is happening along coasts, where half the people of the world live. The source of the pollution can be divided into two categories.

Point source pollution is pollution that can be traced to a single source. This includes industrial waste, of which 5 trillion tons of waste is dumped into waterways annually; and *sewage* pipes, which dump another 3 trillion tons of treated sewage into waterways. New York City alone dumps about 40 million gallons of partially treated sewage into New York rivers every day, which quickly makes its way to the ocean.

Pointless pollution, also called nonpoint pollution, is pollution from many sources. It includes water runoff containing *fertilizers, pesticides, oils,* and solvents that collect along shores, causing high pollution levels.

Municipal solid wastes, sewage *sludge,* and *hazardous wastes* still pour into the oceans. It may be point or pointless pollution, legal or illegal, but on a worldwide basis, the ocean is still used as a toilet bowl. All of this pollution cannot simply be diluted away. Major fisheries are often contaminated, beaches are closed, aquatic ecosystems are destroyed, and species forced into extinction. Up till the mid 1980s, *nuclear waste disposal* meant dumping into the oceans, and ocean dumping of wastes from barges was common.

The Federal Water Pollution Control Act of 1956, which was amended to include the *Clean Water Act* of 1977, and the Ocean Dumping Act of 1972 authorize the EPA to regulate ocean pollution. The *United Nations Environment Programme* works on an international scale at controlling ocean pollution.

ocean thermal energy conversion (OTEC)   OTEC is an experimental method of generating power that uses differences in temperature between the ocean's colder, deeper water and the warmer, more shallow water. OTEC power plants are anchored to the ocean bottom and float at the surface. Japan and the U.S. are currently studying the feasibility of these power plants. These types of plants produce no pollutants, similar to *hydroelectric* and related forms of power generation. (See *Wave power, Moon power*)

oceanic zone ecosystems Marine ecosystems can be divided into the coastal region (*neritic zone*) and the open seas (oceanic zone). The oceanic zone is found beyond the continental shelf, which can protrude hundreds of miles from the shore. Light penetrates only the surface layer of the open sea down to about 600 feet. This is called the *euphotic zone,* where *producers* perform *photosynthesis* and most marine life exists. Beneath this layer is the aphotic zone, where little or no light penetrates.

The ocean receives a constant flow of nutrients from rivers and the water runoff from the surrounding land, but most of these nutrients remain in the coastal region, leaving the open ocean to be one of the least productive environments on the planet.

Where ocean currents carry nutrients out to the open ocean, *food chains* exist. *Phytoplankton,* in the euphotic zone, perform most of the photosynthesis of the sea. *Zooplankton* consume the phytoplankton, which are in turn eaten by many types of small fish or crustaceans such as shrimp. Tuna and other large fish eat these animals and sharks and other predators complete the chain. Many of the larger fish live below the euphotic zone, but visit the surface in search of food.

Organisms that live beneath the euphotic zone and don't venture up to it must depend on the constant downward flow of dead organisms (mainly plankton) as their source of food, meaning they are *scavengers* or *decomposers.*

Oceanic circulation supplies all these deep-sea organisms with oxygen to survive. A few deep-sea *predators* have been discovered in the lower portion of the aphotic zone, and some organisms have been found living on the floor of the oceans, even at depths of 6000 to 8000 meters (3.5 to 5 miles down). These bottom dwellers, called benthic organisms, include unusual species of sea urchins, starfish, and strange tube-dwelling worms. Although they don't live in the water, many birds also play a role in these ecosystems since they feed on animals near the surface.

oceanography The study of marine organisms and their habitats. (See *Marine ecosystems*)

oceans Our planet is often called the water planet, since over 70 percent of the surface is covered with water. Ninety-seven percent of this is found in the oceans. All waters from streams and rivers ultimately end up in the oceans. The oceans play a key role in regulating the planet's climate by storing heat and distributing it through worldwide currents. Oceans help regulate the **water** (hydrologic) **cycle** and play an important role in many other **biogeochemical cycles.** The oceans also store vast quantities of carbon dioxide, which help regulate the overall temperature of the planet. Finally, oceans are home to roughly 250,000 species. (See *Oceanic zone, Wave power, Tide power, Ocean thermal energy conversion*)

Office of Technology Assessment The OTA reports to Congress on the scientific and technological impact of government policies and proposed legislation. It studies science and technology issues for Congress, attempting to resolve conflicting claims, beliefs, and data. This office has substantial influence on the strategic direction the U.S. takes regarding environmental and energy issues.

It was created in 1974 and has a board, which consists of six senators appointed by the president pro tempore, and six representatives appointed by the Speaker. The board elects a director, who serves a six-year term. An advisory council of 10 eminent scientists is called upon regularly for input.

oil Oil began to replace coal during the early 1900s, primarily because of the automobile. Crude oil (also called petroleum) is a thick, black liquid substance composed mainly of **hydrocarbons** with small amounts of other chemicals such as sulfur. It is found deep in the earth's crust or beneath the ocean floor. The oil is found within the pores and cracks of rock.

About 42 percent of the U.S. energy supplies comes from oil. Seventy-seven percent of the world's **oil reserves** are located in the Middle East, with most of it in only five countries: Saudi Arabia, Kuwait, Iran, Iraq, and the United Arab Emirates. About 67 percent is in the hands of **OPEC** nations. The U.S. uses 30 percent of all the oil extracted globally each year but only possesses about 4 percent of the world's reserves. At current levels, the world's oil reserves are expected to last about 50 years. Optimistic projections estimate there might be 100 times more oil still undetected within the earth, but most of it could not be economically extracted with existing technologies.

Oil has a high heat content and burns relatively cleanly if used with proper pollution control devices. Burning oil, as with other fossil fuels, is a major contributor to **greenhouse gases** and some air-pollution problems. Oil tanker spills, such as from the **Exxon Valdez,** have caused localized environmental destruction. (See *Cool energy*)

oil formation   Oil, like other **fossil fuels,** was created from accumulations of dead organisms that were exposed to intense heat and pressure for hundreds of millions of years. While **coal** was created from vegetation, oil is believed to have formed from micro- scopic marine organisms that died and accumulated in the sediment at the bottom of the oceans. As these organisms died, they released small quantities of oil. The sediment gradually became shale, which contained the accumulated oil. Oil deposits were created when geological conditions allowed the oil in the shale to be forced into surrounding layers of sandstone, which soaked up the oil like a sponge. (See *Coal formation*)

oil pollution in the Persian Gulf   Oil tankers in the Persian Gulf normally wash out their holds and dump about two million barrels of oil each year, directly into the gulf. This large volume of oil pollution was dwarfed during the Persian Gulf War of 1991, which produced the largest oil spill in history with four to eight million barrels of oil spilling directly into the gulf waters within a few weeks. Seven hundred wells were blown up by retreat- ing Iraqi forces, most of which were set afire and burned for almost one year, blackening the skies.

Almost all the oil remained within 250 miles of the spill along the Kuwait and Saudi Arabian coastlines, killing wildlife and causing respiratory illness in those living nearby the shore. It is believed to have contaminated some of the groundwater drinking supply. The cleanup, including the cost of putting out all the well fires, ran about 20 billion dollars. (See *Persian Gulf War pollution, Oil spills in U.S., Exxon Valdez*)

oil recovery   Oil is extracted in one of three ways. Primary oil recovery involves drilling a well and pumping out the oil that ac- cumulates at the bottom. Secondary recovery attempts to get the thicker, heavier oil that does not accumulate by the natural force of gravity. In this method, water is injected into an adjacent well in an effort to force remaining oil into the central well. Both primary and secondary recovery extracts only about one-third of the total oil content.

Tertiary or enhanced recovery is expensive and involves pumping steam into the well in an attempt to reduce the viscosity of the heavy oil and get it to flow into the well.

Once removed from the well, the crude oil is usually sent through a pipeline to a refinery, where it is separated by heat into gasoline, heating oil, asphalt, and many other substances. (See *Natural gas extraction, Strip mining*)

oil spills in U.S.   Every day, about 50 oil tankers, each carrying about 450 million gallons of oil, enter U.S. harbors. Between 1978 and 1991, there was an average of 6.3 oil spills per year, dumping 3.4 million gallons into the waters each year. 1991 had the lowest average since 1978, with only 55,000 gallons spilled. Experts believe this reduction might continue, and they associate the smaller number with the passage of the Oil Pollution Act of 1990, which increased the liability of tanker owners and required them to take greater safety precautions. (See *Oil pollution in Persian Gulf*)

old-growth forests   To a forester, old growth means trees that are at their maximum harvest size. To others it refers to a stand of large, old, living trees that have never been harvested (called primary forests). A classic old-growth forest, according to the **Wilderness Society,** contains at least eight big trees per acre, each exceeding 300 years in age or measuring more than 40 inches in diameter. (See *Ancient forests, Douglas fir, Coast Redwood*)

oligotrophic lakes   Refers to deep, clear, cold lakes containing few nutrients to sustain life. (See *Eutrophication, Standing water habitats*)

omnivore   Animals that eat both plants and animals are omnivores. Bears, raccoons, foxes, and most humans are omnivores. (See *Food chain, Energy pyramid, Consumer*)

oology   Pertains to the study of eggs.

ooze   Refers to fine grain **sediment** found at the bottom of deep waters, containing the remains of marine organisms.

OPEC The Organization of Petroleum Exporting Countries was created in 1960 when five of the world's major *oil*-exporting countries got together and formed an oil cartel in an effort to control the global price of oil. There are 13 countries in OPEC, including seven Arab countries: Saudi Arabia, Kuwait, Libya, Algeria, Iraq, Qatar, and the United Arab Emirates; and six non-Arab countries: Iran, Indonesia, Nigeria, Gabon, Ecuador, and Venezuela. These countries control over 60 percent of the world's *oil reserves* and about 30 percent of the current oil market. Differences between the member nations have weakened the cartel in recent years. (See *Oil resources, Oil recovery, Cool energy*)

organic beef Refers to beef that have been raised without antibiotics, growth hormones, or synthetic chemicals of any kind. Very little beef on the market is considered organic or naturally grown beef. There are a few companies that are blazing the trail to raising and selling more natural producing beef; three such companies are: Coleman Natural Beef, Inc., (303) 297-9393, Country Natural Beef, (503) 576-2455, and B3R Country Meats, (817) 937-3668. (See *Factory farms, Organic farming*)

organic farming Organic farming is a new term used to define old farming practices that don't depend on synthetic *insecticides* and *fertilizers.* Manual or mechanical labor is used to remove weeds. Mixed crops are planted to avoid pest infestations. *Organic fertilizer* (usually animal manure) is used to enrich the soil, and *crop rotation* is used to stabilize these nutrients. The amount of energy needed to run an organic farm is much less than that used on highly mechanized, modern *monocultures,* but the yields are somewhat lower.

Food grown on farms that avoid pesticides and fertilizers are now often called "organically grown" foods and have recently gained in popularity. Many countries have seen dramatic increases in the popularity of organic farming, including Canada, Australia, Israel, and New Zealand. Japan alone has over 20,000 organic farms in operation. Increased public acceptance of organically grown foods is likely to drive the price down by getting more and more farms involved and reduce the amount of harmful pesticides and fertilizers applied each year.

Some of the practices used with organic farming, such as *crop rotation* and resistant varieties of crops, are used on large, mechanized farms, as an important part of an *integrated pest man-*

agement plan. (See *Sustainable agriculture, Natural insecticides, Biological control, Biological pesticides*)

organic fertilizer Organic fertilizer, like any fertilizer, is used to enrich the soil by adding nutrients required for plant growth. Organic fertilizer comes from **organic matter.** There are three types of organic fertilizer. 1) **Animal manure** is composed of animal dung and urine, usually from cattle, sheep, horses, or poultry. Some **less developed countries** use human dung, called **night soil**, as an organic fertilizer. 2) **Green manure** consists of any plants that are plowed into the soil; and 3) **compost.** (See *Organic farming*)

organic food supermarkets Organically grown produce and **organic beef** are sold in about 6,000 small specialty food markets, but recently a number of large organic food supermarkets have appeared. There are about 50 organic (out of roughly 30,000) food supermarkets in the U.S. Some people, concerned about their health and the environment, prefer to eat produce that hasn't been grown with synthetic **fertilizers** and **pesticides,** and meats that haven't been raised on synthetic hormones and antibiotics. Bread & Circus, Whole Foods Market, and Alfalfa's are just a few of the organic food supermarket chains. (See *Biological pesticides, Natural insecticides, Biological control, Organic fertilizers*)

organic labelling Foods grown on organic farms are becoming increasingly popular and have resulted in a flurry of marketing and advertising claims. A few groups, such as the Organic Crop Improvement Association (OCI), which is international, and the Natural Organic Farmers Association (NOFA), which is regional, have established reliable certification programs assuring that products have truly been organically grown. There are many standards, but the most important is that the land the products are grown on be free of chemical **pesticides.** The federal government has also gotten involved since the 1990 Farm Bill established some of its own standards. NOFA is located in Barre, Massachusetts, and OCI is located in Belle Fontaine, Ohio. (See *Organic farming*)

organic matter Refers to substances that compose or are derived from living organisms. All organic matter contains carbon as the essential element.

organophosphates (OPs) Organophospates is one of four major categories of synthetic **insecticides.** OPs, along with **carbamates,** are considered "soft pesticides" since they break down into harmless chemicals shortly after application, usually 1 to 12 weeks. OPs kill by effecting the normal functioning of nerve cells. Even though they are less persistent than **chlorinated hydrocarbons,** they are usually far more toxic to humans, meaning they are especially dangerous to those who apply the chemicals or are in the vicinity of the application. Since OPs pose a significant health risk, many have been replaced with carbamates.

Some examples of OPs are malathion, often used to control mosquitoes, and diazinon, sold as Spectracide, commonly used for pest control in household gardens. Most flea collars and pest strips also contain OPs. (See *Pesticides, Organic fertilizer*)

outdoor recreation The three most popular outdoor recreational activities in the U.S. are **bicycling,** fishing, and camping, each with over 20 million participants annually. Other popular outdoor activities include jogging, hunting, golf, hiking, tennis, and skiing, all with between 7 and 15 million annual participants. (See *Outdoor recreation, motorized vehicles; Bicycling, best cities for*)

outdoor recreation, motorized vehicles To some people, **outdoor recreation** means motorized vehicles such as all-terrain vehicles and skimobiles. Nonvehicular outdoor recreation causes some damage to the environment, but motorized vehicles inflict much greater damage. For example, all-terrain vehicles kill the vegetation along trails and usually result in erosion and substantial damage to the area ecosystem. Since public land is owned and paid for by everyone, there is a great deal of controversy about this type of land use.

Outdoor Writers Association of America (OWAA) The OWAA is an organization composed of professional writers who report on interests related to the outdoors. Topics include fishing, wildlife photography, and environmental issues. Members of OWAA are not only professional journalists but many have degrees in wildlife management, environmental science, forestry, and allied fields. They share information through conferences, newsletters, and environmental committees. Write to 2017 Cato Ave., Suite 101, State College, PA 16801.

outgassing   Outgassing is a form of evaporation that occurs when unstable molecules are released into the air. A good example is the release of formaldehyde from building materials and products such as particleboard, paneling, carpeting, and furniture. Outgassing occurs when **formaldehyde** escapes from the glue found in these products. Intense outgassing occurs the first few weeks of the product's life, and then levels off to low-level outgassing that can occur for years. Outgassing may be a factor in **sick-building syndrome** and **building-related illness.** (See *Baubiologie, Healthy homes, Volatile organic compounds*)

Outward Bound USA   Outward Bound is an adventure-based educational program designed to teach leadership training through challenging activities in a wilderness setting. Their focus is to help individuals learn about themselves. The instructors are professional outdoor educators. Some of the courses include: whitewater rafting, canoe expeditioning, sailing, sea kayaking, backpacking, horsetrailing, and trekking. Students must be fourteen or older. Write to 384 Field Point Road, Greenwich, CT 06830. (See *Primitive technologies*)

overburden   Refers to the material found above a vein of coal that must be removed. (See *Strip mining*)

overgrazing   Refers to animals feeding (grazing) on vegetation at a rate exceeding the vegetation's ability to recover, resulting in damage to the ecosystem. (See *Erosion, Desertification*)

overpopulation   Overpopulation means there are more individuals in a population than the resources can provide for. The existence of too many rabbits, with not enough food, water, or shelter, results in overpopulation, death, and a decline in the rabbit population. More specifically, the term overpopulation is used when the **carrying capacity** of an area is exceeded, and the survival of individuals within the population is threatened. This is followed by a decline in the population by natural forces such as starvation or disease.

When speaking about human populations, we include not only "survival," but also the health and welfare of the individuals when determining whether overpopulation exists. (See *Population, Doubling-time in human populations, Tolerance range, Limiting factor*)

overstory   The uppermost layer of a forest community, which is formed by the highest trees. The highest **canopy** forms the overstory. (See *Tropical rain forest*)

oviparous   Refers to egg-laying animals. Animals that produce eggs that develop and hatch outside of the maternal parent are oviparous and include birds, amphibians, and most insects. (See *Ovoviparous, Viviparous*)

ovoviparous   Refers to animals that produce eggs that develop within the maternal parent but hatch either inside or immediately upon leaving the parent's body. Sharks, lizards, and some insects and snakes are ovoviparous. (See *Oviparous, Viviparous*)

ozone   Ozone is associated with two distinct environmental problems. The first involves ozone that occurs naturally in the upper atmosphere, where it is called stratospheric ozone, or the ozone layer. **Ozone depletion** of this layer appears to be one of the most pressing environmental problems of our times.

   The other involves human activities that produce ozone, usually referred to as ground-level ozone. Burning **fossil fuels** releases primary air pollutants, which react with one another to create new chemicals called **secondary air pollutants**. Ground-level ozone is one of these secondary pollutants. It irritates the eyes and causes respiratory problems. It is also a **greenhouse gas** and plays a minor role in contributing to **global warming**.

ozone depletion   Depletion of the **ozone** layer (stratospheric ozone) has become one of the most pressing environmental concerns of these times. Since ozone absorbs the sun's ultraviolet (UV) light, the ozone layer is a vital protective shield for life on earth. UV light is harmful to most organisms, including humans. It disrupts DNA and can cause genetic defects in both plants and animals. It causes skin cancers and eye cataracts in humans and reduces yields in food crops such as corn and rice. A 1 percent loss of the ozone layer is expected to result in a 5 percent rise in the incidence of skin cancer.

   During the 1980s a hole containing 50 percent less ozone than expected was found in the ozone layer over the Antarctic during September and October of each year. On some occasions the hole was the size of the entire continental U.S. The UV radiation reaching the surface increased about 20 percent during these periods. Scientists later discovered a hole over the Arctic, missing 20 percent of its usual complement of ozone. After a few months, this hole breaks up and becomes dispersed over much of Europe and North America. The 1988 estimates have the overall annual loss of ozone over most of North America, Europe, and Asia at about 3 percent when compared with 1969.

Ozone depletion is caused by *ozone-depleting gases* that accumulate in certain regions of the stratosphere during certain times of year due to weather conditions. Since these gases take a long time to work their way into the atmosphere and remain active for many years, there will be lag time of decades, if not centuries, before any remedies we establish today take effect.

ozone-depleting gases *Ozone depletion* of the ozone layer is caused by accumulations of gases introduced into the air almost exclusively by human activities. The primary cause of ozone depletion is a family of gases called *CFCs (chlorofluorocarbons)* used as refrigerants and in aerosols. Other contributors are *halons,* and a few solvents such as carbon tetrachloride and methylchloroform.

Under normal conditions, ozone absorbs ultraviolet light and breaks down into a molecule of oxygen and a free oxygen atom. These two parts naturally recombine to reform ozone and perpetuate the existence of the ozone layer. Chlorine from CFCs, however, interferes with this process. The chlorine reacts with ozone, producing an oxygen molecule, but no free oxygen atoms. (They are tied up with the chlorine.) This means the ozone cannot continually reform, and the ozone layer becomes depleted. This forms the well-known hole in the ozone layer, allowing excess UV light to reach the earth's surface.

ozone, stratospheric Stratospheric ozone forms the *ozone* layer, which protects life on our planet from harmful ultraviolet light (UV) from the sun. Ozone is composed of three oxygen atoms, which soak up the UV light from the sun. As it absorbs sunlight, it breaks into smaller pieces. These smaller pieces usually consist of one oxygen molecule and one free oxygen atom. Oxygen molecules and free oxygen atoms readily recombine to reform ozone, so the process perpetuates itself. (See *Ozone depletion, CFC*)

palynology   The study of pollen and spores.

Pangaea   The name of the super continent believed to have existed about 240 million years ago, which began to break up into the existing continents about 200 million years ago.

panspermia   A theory that life on earth was initiated by the introduction of life from somewhere else in the universe (i.e., extra-terrestrial).

paper recycling   Landfills are not "filling-up" with disposable diapers or plastics nearly as much as with paper. Recycling paper products helps reduce the ***municipal solid waste*** dilemma, but there is a great deal of confusion about what is and isn't recycled paper. Saying a paper product is recycled tells only a portion of the story. More important is the percentage of the product that is recycled from "post-consumer" waste. In other words, how much of the product was already used in another consumer product, and therefore truly recycled?

The term "post-industrial" waste means the material was never in the consumers' hands, but recycled from some industrial process. For example, wood chips or sawdust collected from a timber mill could be collected and called recycled, post-industrial waste. This is better than simply disposing of the material, but not truly recycled.

Whenever you purchase recycled paper products of any kind, look for the amount of post-consumer waste, post-industrial waste and virgin materials included. For example, some environmentally conscious companies are producing paper products marked as recycled from 80% post-consumer waste, 10% post-industrial waste, and 10% virgin materials. If it is not clearly marked, assume the worst. (See *Recycling, Green marketing, Seals of approval*)

parasite   When one organism lives on or in another organism for at least a portion of its lifetime to gain nourishment, it is called a parasite and the relationship is parasitism. The organism gaining nourishment is the parasite, and the organism providing nourishment is the "host." The parasite obviously gains from this relationship, but the host is usually harmed to some degree but not killed, since killing the host eliminates the food supply for the parasite.

Numerous insects such as lice and fleas are **ectoparasites** that use humans as well as other animals as hosts. The tapeworm, some protozoans, and many bacteria are **endoparasites** in humans and other animals. The mistletoe is an example of a flowering plant that parasitizes trees. (See *Symbiotic relationships, Commensalism, Mutualism, Parasitoidism, Biological control*)

parasitoidism   This is a **symbiotic relationship** that is a cross between **parasitism** (with its parasite/host relationship) and a **predator-prey relationship.** Parasitoids lay eggs in a host. When the young hatch, they feed upon the host, either internally or externally. Once mature, the adult parasitoid exists the body of the host. The host usually dies as a result of the invasion. Most parasitoids are insects from either the Hymenoptera (wasps) or Diptera (flies) orders. Parasitoids are sometimes used as an alternative to synthetic **insecticides.** (See *Biological control, Integrated pest management*)

parent rock   **Soil** is formed when solid rock breaks down to small particles. The type of solid rock (along with the climate and animal life that exists in the region) dictates the type of soil that will form. The rock from which a soil is formed is called the parent rock or parent material.

The parent rock is broken down by different types of weathering. Physical weathering involves the effect of temperature changes (freezing and thawing); mechanical abrasion is caused by rock rubbing against rock, pressure exerted by plant roots, or a glacier moving over it. Chemical weathering is caused by chemical contact from exposure to air, water, or products of plants and animals. Soil continues to be formed as the particles become smaller and biological processes become involved. (See *Soil particles, Soil horizon*)

parthenogenesis   Refers to the creation of an individual with an unfertilized egg, which is a form of **asexual** reproduction. Some insects, such as aphids, can reproduce parthenogenically.

particulate matter   Particulate matter, also called suspended particulate matter (SPM), is one of five types of primary pollutants contributing to **air pollution.** It refers to any tiny solid particle, such as dust (which is fine **soil**) soot, which is fine carbon particles, and any other fine particles dispersed from **pesticides, asbes-**

tos, or thousands of other products. Many of these are simply an annoyance, but others are *toxic.* Particulate matter often attracts and carries chemicals through the air such as dust carrying sulfuric acid. It is the most noticeable type of air pollution since it is readily visible.

Particulate matter is reduced by air pollution control devices in industry and electric power plants. Scrubbers, filters, separators, and precipitators are either built into new construction, or often retrofitted into old facilities. Automobiles also have emission control devices to reduce particulates.

Large numbers of wood-burning stoves and fireplaces release enough particulates to produce a symptom called *brown cloud.* Many municipalities regulate the number and efficiency of wood-burning stoves to reduce these emissions.

parts per million (ppm) Ppm is the unit of measurement used when quantifying the minute amounts of a gas in the air. For example, concentrations of .25 ppm of *formaldehyde* in the air (given off by office furniture and carpeting) is not irritating to most adults, but .5 ppm can cause burning of the eyes and respiratory irritation in many individuals. (See *Outgassing*)

passive smoke See *Environmental tobacco smoke (ETS).*

passive solar-heating systems Passive solar-heating systems simply absorb the sun's energy in the form of heat for immediate use. Insulated windows, for example, allow the radiant energy from the sun in, but don't let the resulting heat out. Specially designed and oriented walls made of concrete, adobe, brick, or stone absorb the sun's energy and gradually release the heat into the building after the sun sets.

Homes or other buildings designed to use solar heat dramatically reduce the need for other energy sources. There are about 300,000 homes and 17,000 commercial buildings using passive solar heating with *super-insulation* (to keep the heat in) in the U.S. These buildings get about 35 percent of their energy needs directly from the sun but most require backup conventional heating systems. The technology currently exists, however, to supply about 80 percent of a new building's heating needs with passive solar facilities. Solar heat can also provide hot water for a home with the use of passive solar water heating systems that pass water through panels that are heated by the sun.

A good quality passive solar-heating system in a **super-insulated** home is believed to be the least expensive way to heat a home in regions where there is enough sunlight. Even though it raises the cost of a new home by 5 to 10 percent, it reduces the overall operating costs of the home by 30 to 40 percent. (See *Solar power, Solar thermal power plants*)

patchwork clear-cutting See *Clear-cutting.*

pathogen Refers to a disease-producing organism.

payroll deduction, environmental Payroll deductions are often used as a vehicle to donate money toward charitable organizations. Earth Share, of Washington DC, offers a payroll deduction plan that allows people to make charitable donations exclusively toward environmental organizations Call (800) 875-3863. (See *Money market funds, environmental*)

PCB (polychlorinated biphenyls) *See* Volatile organic compounds.

PCV valve The positive crankcase ventilation (PCV) valve is required on automobiles sold in the U.S. It reduces **hydrocarbon** emissions, a major component of **air pollution.**

peat Refers to a rich, fertile soil composed of at least 50 percent **organic** matter. Contains more organic matter than **muck.** (See *Soil types, Humus, Compost, Ooze*)

pelagic marine ecosystems **Marine ecosystems** can be divided into two major types: benthic marine ecosystems are found at the bottom of the oceans and seas, whereas pelagic marine ecosystems are found in the open bodies of water. pelagic ecosystems are divided into regions: **neritic** (coastal) and **oceanic** (open seas).

The neritic region is found above the continental shelf. It begins at the continental shores and extends out to sea usually several miles and reaches a depth of about 600 feet. This zone takes up only 10 percent of all the oceans but contains 90 percent of all plants and animals. Oceanic ecosystems are found beyond the continental shelf and are considered relatively nonproductive when compared with the coastal regions and most terrestrial ecosystems (**biomes**).

percolation   Refers to the downward flow of water through soil or solid waste. For example, rainwater percolates through a **landfill,** producing **leachate.**

peregrine falcon   Believed to be the world's fastest animal, it can attain speeds of 200 miles per hour in an aerial dive. This bird almost became **extinct** during the late 1960s and early 1970s due to the use of the insecticide **DDT.** During the mid 1970s, the birds made a recovery thanks to restricted use of DDT and efforts from organizations such as the Peregrine Fund and the National Park Service. (See *Endangered Species Act*)

perennial plants   Refers to plants that live more than three years and have specific growth periods each year. Most vines, shrubs, and trees are perennials. (See *Annual plants*)

permafrost   Refers to any part of the earth's solid surface that remains frozen for two or more years. (See *Tundra*)

Persian Gulf War pollution   Although brief, the Persian Gulf War of 1991 produced an environmental disaster. Almost 700 oil well fires were set in Kuwait by the retreating Iraqi forces and another 100 wells produced a multimillion-gallon oil spill in the gulf. Severe **air pollution** in the skies over Kuwait and Saudi Arabia, caused by the fires, forced about 10 million people to breathe the polluted air during this time. Much of the oil contains sulfur, which, when burned, produces sulfur dioxide, contributing to **acid rain.** The oil that spilled into the gulf caused immediate damage by killing thousands of birds and other wildlife, but the long-term effects on the entire ecosystem can't be determined for many years. (See *Eco-terrorism, Oil pollution in the Persian Gulf*)

persistence (population)   See *Population stability.*

persistent pesticides   Refers to **pesticides** that last for many years, so they need not be applied often. Persistent pesticides are also called "hard pesticides" and pose significant health risks. (See *Chlorinated hydrocarbons, Bioaccumulation, Biological amplification, Pesticide dangers*)

pest   Refers to any organism found in a location where humans prefer it not to be. Although the term can be used for both animals and plants, "weed" is often used for plant pests. (See *Pesticide*)

pesticide   Pesticide is a generic name for poisons that kill unwanted organisms. Pesticides have historically saved lives by protecting people from insect-transmitted diseases and increasing the food supply. Pesticides were once thought to be a panacea, but many **pesticide dangers** have become evident over the past few decades.

About 2 million tons of pesticides are used each year, approximately 1 pound for every person on earth. Pesticides can be divided by the type of pest being targeted for destruction. **Insecticides** kill insects, **herbicides** kill weeds, fungicides kill fungus, rodenticides kill rodents, etc. In the U.S., about two-thirds of all pesticides used are herbicides, 23 percent insecticides, and 11 percent fungicides. The majority of pesticides are used on only four crops: corn, cotton, wheat, and soybean. Twenty percent of all pesticides in the U.S. are not applied to crops, but to lawns, gardens, golf courses, and other nonfarming lands. (See *Biological control, Biological pesticides, Natural insecticides, Integrated pest management*)

pesticide dangers   Each year, roughly 4 billion pounds of **pesticides** are used worldwide. There are many problems associated with pesticides, including the fact that most do not kill only the targeted pest. Most are harmful to many organisms, including humans. For this reason, many people prefer to use the term "biocide," since they don't just kill pests.

Numerous studies have found birds, mammals, and fish to be poisoned by pesticides either by direct contact or from **biological amplification.** Pesticides have reduced populations and driven some species to the brink of extinction. Many pesticides leach down through the soil, contaminating **groundwater,** which supplies 50 percent of our drinking water. Pesticides have been found in groundwater in 26 states.

There are roughly 50,000 cases of direct pesticide poisoning reported in the U.S. annually. Possibly more serious, however, are the long-term effects of pesticides in our water and on our foods. Studies have shown that many of these substances increase the risk of cancer, birth defects, and many chronic illnesses.

Many pesticides are becoming ineffective since pests often become resistant to the poison through the process of **natural selection.** The more resistant they become, the more chemicals must be used, and the more harm done to the environment and to us. Because of this resistance, the quantity of pesticide required to kill is often increased by a factor of 100 after a few years of use. In the last 40 years, the amount of pesticides used has increased over ten times, but the amount of crops lost to the pests has still continued to rise from 7 percent to 13 percent.

The use of pesticides can turn nonpests into pests. Some insects, normally controlled by natural ***parasites*** and ***predators,*** increase in numbers dramatically when pesticides kill their predators and parasites. Increased numbers often turn an otherwise harmless insect into a pest.

Pesticides end up everywhere: in groundwater, rivers and streams, ponds, lakes and oceans, in the air, in the soil, and on and in plants and animals, including humans. Some pesticides have even been found in breast milk. There are two alternatives available. First, on an individual basis, foods grown on organic farms, which avoid the use of pesticides, are often available. On a larger scale, farmers have alternative and proven methods of protecting their crops without extensive use of pesticides, including ***integrated pest management (IPM)*** and ***biological control methods.*** (See *Organic farming, Pesticide residues on food, DDT, Carson, Rachel; Breast milk and toxins*)

pesticide regulations The threat posed by pesticides was first brought to the public's attention by ***Rachel Carson*** in her classic 1962 book, *Silent Spring.* The dangers have been well founded and still exist today. The National Academy of Sciences, the General Accounting Office, and most ***environmentalists*** believe the law created to protect U.S. citizens from dangerous pesticides is poorly enforced and poses a threat to the American public. The law they were referring to is the ***Federal Insecticide, Fungicide, and Rodenticide Act of 1972,*** called FIFRA for short. The ***Environmental Protection Agency (EPA)*** is responsible for carrying out this act.

FIFRA gave the EPA the responsibility of analyzing the potential dangers of about 700 active ingredients in pesticides used today but registered before 1972, when registration procedures were improved. The act stated that testing these substances was to be completed by 1975. By 1989, fewer than 200 had been studied. The EPA was granted an extension to 1997, but is already projecting the task won't be completed until after 2000.

Current procedures let the EPA leave chemicals that have never been adequately tested and may pose a threat to the public health on the market until they are tested. Amendments to the Act in 1988 have made it easier for the EPA to ban a chemical if necessary, but this has rarely been done.

pesticide residues on food   The Food and Drug Administration (FDA) is responsible for testing food products for pesticide residues. Everyone agrees pesticide residues exist on most foods, including fruits, vegetables, and grains, but the question is: "How safe are they?" The FDA believes the levels and types of pesticides found on our foods are safe; the National Academy of Sciences isn't sure, and many environmental groups are positive they are unsafe. Some recent studies have found that 67 of the 300 pesticides commonly used on food products can cause cancer in laboratory animals.

The FDA tests only 1 percent of the food supply for pesticide residues that exceed approved levels. Over half of the foods that do get tested have pesticide residues. As long as the amount found doesn't exceed accepted tolerance levels, these residues are legal.

Pesticides banned in the U.S. are still exported to other countries. Many of these banned pesticides are used in other countries on crops that are then exported back to the U.S. Therefore, pesticides banned by the U.S. make their way back to our tables on imported foods. This has been called the **circle of poison.** FDA testing consistently shows higher pesticide residues on imported fruits and vegetables compared with domestically grown produce.

The U.S. regulations also allows pesticides to contain large amounts of impurities. These impurities can even include the active ingredients of "banned" pesticides. Some pesticides sold today contain as much as 15% **DDT,** a pesticide banned decades ago. (See *Fruit waxing, Food irradiation*)

PET   PET is an acronym for poly-ethylene terephthalate, which is the **plastic** used to produce rigid containers such as carbonated soda bottles. It makes up only about 7 percent of all the plastics produced in the U.S. but is one of the few plastics commonly recycled. For **plastic recycling** purposes these products are stamped with the number "1" surrounded by a triangle made of arrows. (See *Plastic pollution*)

petrochemicals   Petrochemicals are substances obtained during crude **oil** refinement. They are used as raw materials in the manufacture of products such as *plastics,* synthetic fibers, paints, and *pesticides.* (See *Oil recovery*)

petroleum See *Oil.*

pheromone Refers to a substance produced by an organism to communicate information to other members of the same species. For example, individuals of some species of ants and aphids, if attacked by a predator, secrete an alarm pheromone, warning other members of the colony. (See *Sex attractants*)

philosophers, environmental Refers to men and women who have helped us develop a consciousness and awareness about how we are part of the whole and must fit in instead of take over. In the early 1800s, Ralph Waldo Emerson's essay "Nature," and later in that century, Henry David Thoreau's *Walden Pond,* helped give birth to the environmental movement. *John Muir*'s writings and his establishment of the *Sierra Club* in 1890 helped strengthen environmental awareness.

In the early 1900s, *Aldo Leopold* helped combine philosophy and science in his writings and studies. He philosophized in *A Sand County Almanac* and proposed scientific hypotheses in the Bulletin of the American Game Association. More modern-day environmental philosophers are often scientist turned philosopher. The most well known is *Rachel Carson,* who wrote the classic *Silent Spring* in the early 1960s, warning everyone about *pesticide dangers,* which many consider the catalyst for today's environmental movement. (See *Ethics, environmental; Environmentalist*)

phosphorus cycle Phosphorus is an essential *nutrient* used by organisms to build DNA and cell membranes. *Inorganic* phosphorus is found in phosphate rocks. Weathering and erosion break down the rock, allowing phosphorus to dissolve in water and become available for green plants to absorb through their roots. The phosphorus is passed along *food chains* to animals. When plants and animals die, the organic molecules containing phosphorus are decomposed back into simple inorganic phosphorus to be used by other green plants. (See *Biogeochemical cycles*)

photochemical smog Burning *fossil fuels* releases five primary air pollutants. Many of these primary pollutants react with each in the presence of sunlight, creating *secondary air pollu-*

tants. Pollution caused by these secondary pollutants is called photochemical smog.

Ozone is a major component of photochemical smog. Nitrogen compounds (a primary pollutant) react in the presence of sunlight to create ground-level ozone (a secondary pollutant). Ozone damages chlorophyll in plants and lung tissues in animals. After the ground-level ozone has formed, it combines with other primary pollutants (such as hydrocarbons) to form new substances called PANs for short (peroxyacyl nitrates), which causes severe eye irritation. Photochemical smog is intensified by certain climatic and geographical factors in a process called *thermal inversion.*

photosynthesis   Radiant energy from the sun is captured by green plants and converted into chemical energy during photosynthesis. The chemical energy is in the form of sugar (glucose) molecules. This energy is then released during *respiration* and used by plants and animals (by way of *food chains*) to survive. Photosynthesis uses *carbon dioxide* and *water* to form this sugar and releases oxygen in the process. (See *Carbon cycle, Biogeochemical cycles*)

photosynthetically active radiation (PAR)   Refers to those wavelengths of light that are capable of driving the process of *photosynthesis.*

photovoltaic cells   Photovoltaic cells, also called solar cells, convert solar energy directly into electricity. Photovoltaics were first created in 1954 by Bell Laboratories and became well known years later in solar calculators. Advances in technology now enable photovoltaic cells to meet the electricity needs for 6,000 villages in India and 10,000 homes in the U.S. These U.S. homes are in remote areas such as Alaska, where it is too expensive to tap into the electric power line grid.

These cells contain purified silicon (from sand) and trace substances (such as cadmium sulfide) that conduct small amounts of electricity when struck by sunlight. Since each cell produces a small amount of power, they must be linked together to produce the power requirements of a home. Large surface areas (many square miles) of these cells can act as solar power plants. It is estimated that 25 percent of the world's energy needs and 50 percent of the U.S.'s needs could be supplied by *solar power* plants within 50 years.

Unlike fossil fuels, this power produces little air and water pollution and no carbon dioxide to add to the *greenhouse effect.* At present, photovoltaic cells are expensive and not cost-competitive, but this is expected to change within the next few years due to new technologies and increased use.

Even though photovoltaic power is considered one of the best bets for supplying the world's future energy needs, U.S. research in this field dropped by 75 percent during a 10-year period beginning in the early 1980s. This has resulted in the U.S. losing 50 percent of its market share in the solar cell world market, while the Japanese doubled their market share. (See *Solar thermal power, Solar ponds, Moon power*)

phreatophyte Refers to a plant that obtains moisture from **groundwater** deep beneath the surface. For example, the mesquite plant can absorb moisture 18 meters (almost 60 feet) down.

phytoplankton Phytoplankton are free-floating microscopic plants consisting of green **algae,** desmids, and blue-green algae. Phytoplankton act as **producers** in many **aquatic ecosystems.** (See *Plankton, Zooplankton*)

phytoremediation Refers to the use of plants to remove or detoxify the environment. (See *Remediation of hazardous waste, Bioremediation, Hyperaccumulators*)

pioneer community The first organisms to establish themselves in an area create the pioneer **community.** This is the first stage of **succession.** The pioneers prepare the environment for the latter stages. **Lichens** are often the first organisms to establish themselves in a pioneer community.

PIRG, U.S. U.S. PIRG (United States Public Interest Research Group) is a national lobbying office that represents state PIRGs around the country. Its purpose is to perform research and lobby for national environmental and consumer protection legislation. The organization focuses on strengthening the **Clean Water Act,** supporting **recycling** policies, and alerting consumers regarding unsafe products, and supports new product safety laws. Write to 215 Pennsylvania Avenue SE, Washington DC 20003. (See *Pesticide dangers, Pesticide residues on foods, Fruit waxing*)

plankton Plankton are microscopic organisms found, usually in large numbers, in slow-moving or standing bodies of water such as oceans, estuaries, lakes, and ponds. Plankton is divided into plant plankton called **phytoplankton,** which include floating green algae, desmids, and blue-green algae, and animal plankton called **zooplankton,** which include crustaceans, rotifers, and protozoans, most of which can move about in the water.

Food chains of many **aquatic ecosystems** begin with phytoplankton, which are the **producers.** They are eaten by the zooplankton, which are the primary **consumers,** and provide a food source for higher organisms such as fish.

plant-oil fuels   Plant-oil fuel is one of three sources of **biomass energy,** in which plants or animals are used to produce useful energy. Some plants, such as sunflower, soybean, and rapeseed, produce natural oils within their seeds. These oils have been used as a replacement for diesel fuel in the past and hold some promise for the future. Plant oils must be refined so they don't foul the machinery they power. Rapeseed can be grown off-season so it doesn't compete with food crops. Some forms of algae also produce natural oils and are being researched for their potential use. Plant-oil fuels are not currently cost-competitive with other fuels. (See *Sunflower oil, diesel fuel*)

plastic   Plastic products are everywhere and make up about 25 percent of the U.S.'s **municipal solid waste** and take up about 8 percent of the space in **landfills.** With few exceptions, plastics take so long to decompose, they are considered nonbiodegradable. Even though over 200 million pounds of plastics are recycled annually, this is less than 5 percent of the total produced. Estimates indicate this number can be ten times greater by the year 2000.

The most common types of plastics are as follows: **HDPE** (high density polyethylene) is used in rigid containers such as those that hold milk or motor oil. **PET** (polyethylene) is also used in rigid containers such as carbonated soda bottles. **PVC** (polyvinyl chloride) is used in construction and plumbing. **LDPE** (low density polyethylene) is used for films and bags. **PP** (polypropylene) is used in snack food packages and **disposable diapers,** plus numerous other products. **PS** (polystyrene) is the plastic foam used for cups and food containers. Finally, there are many miscellaneous plastics used for a variety of purposes such as squeezable bottles. (See *Plastic recycling*)

plastic pollution   Thousands of tons of **plastic** products that aren't recycled or properly disposed of end up floating in bodies of water or littering the land. This plastic might be dumped overboard from commercial vessels, but it also comes from sewer overflows, sloppy garbage handling, and people who litter. Since plastics don't readily decompose, they remain intact in the environment, acting as booby traps for many types of organisms, especially in the oceans. Turtles mistake plastic baggies for jellyfish, eat them, and die from blocked intestines. Birds swallow bits of plastic foam and choke. Animals often become entangled in plastic debris and either drown or starve. (See *Plastic recycling, Cruise-ship pollution*)

Plastic Pollution Research and Control Act
This act was signed into law in 1988 and prohibits ocean dumping of plastic waste. (See *Cruise-ship pollution*)

plastic recycling   The great emphasis recently placed on "biodegradability" has caused confusion about *plastics.* Most plastics were never designed to break down. Plastic waste in *landfills* may never break down, and plastics that litter the water and land will take hundreds of years—if ever—to decompose. Most waste management programs urge that plastics be collected and recycled, something many plastics do quite well. Many parts of the country collect plastics for *recycling* into new products, and new ones are being established all the time.

HDPE and *PET* are both commonly recycled. Scientists, environmentalists, and the plastics industry are currently looking for ways to recycle *PS* products such as *styropeanuts.*

Plastic products that are not likely to get collected, such as plastic fishing gear, or pose a special risk—such as six-pack bottle holder rings, which choke marine animals—can use special degradable forms of plastic currently being tested. These products are supposed to decompose by photodegradation, in which sunlight and microbes deteriorate the product.

Plastic recycling reduces the volume of plastics going into landfills and incinerators, reduces the amount of petroleum needed for their manufacture, and reduces *plastic pollution.* (See *Source reduction*)

plutonium pit disposal   Now that the cold war is over, the U.S. must dispose of over 40,000 nuclear warheads. The plutonium pit is that part of a nuclear bomb that "ignites" the thermonuclear blast. Plutonium remains deadly for over 25,000 years and is considered the most dangerous substance on earth. Grains much smaller than sand can and probably will cause cancer. These plutonium pits are now temporarily stored in concrete and steel bunkers in the

Department of Energy's Pantex Plant in Amarillo, Texas. State officials there are concerned about the **NIMEY** and the **NIMBY syndromes** and about the potential dangers of the facility. (See *Nuclear waste disposal, Hanford Nuclear Reservation*)

poaching   Refers to the illegal capture of wildlife, such as those animals classified as **threatened species** or **endangered species.** International attempts to regulate poaching and illegal trade involving wildlife are carried out by **CITES** and its member nations. The **ivory trade** is a typical example of how poaching has affected wildlife populations, drastically reducing the remaining numbers of the African elephant. The black rhinoceros population, also in Africa, dwindled from about 65,000 in 1976 to only 10,000 a few years later as poachers slaughtered them for their two horns, which are worth tens of thousands of dollars on the black market as a reputed aphrodisiac. (See *Endangered Species Act*)

podzol   See *Soil types.*

point source pollution   Point source pollution refers to pollution that comes from a single, identifiable source such as a manufacturing plant's **effluent** pipe that dumps its industrial waste water into a stream. (See *Pointless pollution, Industrial water pollution*)

pointless pollution   Pointless pollution (also called non-point pollution) does not originate from any one source, but from many unrelated sources. This conglomeration of pollutants often occurs when rain water collects residues of automobile gas and oil on roads, **fertilizers** and **pesticides** from farms and lawns, and animal wastes from feed lots. The water containing pointless pollution is collected and concentrated in drainpipes and drainage ditches and runs directly into bodies of water. It may also be diverted into bodies of water when municipal **sewage treatment** systems overflow.

Coastal areas are the most significantly affected, often resulting in closed beaches and contaminated fishing grounds and shellfish beds. (See *Estuary and coastal wetlands destruction*)

polar grassland   See *Tundra.*

pollination   The fertilization of plants that produce seeds.

pollution   Pollution refers to a negative change in the quality of some part of our **biosphere** or aspect of our lives. These changes, if left unchecked, can cause annoyance, illness, death, or even the extinction of a species.

Pollution is often identified by the part of the planet harmed, as with **air pollution, water pollution,** and **soil** pollution. Pollution can also be categorized by its source, as with **pointless** and **point source pollution** or—more specifically—as with **Persian Gulf War pollution** or **cruise-ship pollution.** Finally, pollution can be identified by the actual product or cause of the pollution, as with **plastic pollution** and **noise pollution.**

Pollution problems were documented as far back as the early Roman empire, but the magnitude of these problems has catapulted in recent decades for two major reasons: human population growth and the proliferation of manufactured products. The number of people and the number of products we manufacture, both of which often occur in dense **urban areas,** have increased and concentrated our wastes—including **sewage, municipal solid waste,** manufacturing by-products, and auto emissions.

pollution police   Almost every state's attorney general's office has some type of environmental-offense unit. Recently, some states have created specially trained environmental crime squads that identify and investigate environmental crimes and enforce environmental laws, much like a felony would be handled. California and New Jersey each have these pollution police squads. In New Jersey, these squads consists of local police officers, sheriffs, and environmental-protection officials who aggressively enforce environmental regulations. (See *Law and the environment*)

population   A population is a group of individuals of the same species living within a specific area. All populations in an area make up a **community.** The community of organisms along with the nonliving parts of the environment such as the soil and water, and factors such as climate, make up an **ecosystem.** (See *R-strategists, K-strategists, Population dynamics, Population stability, Doubling-time in human populations, Carrying capacity, Thomas Malthus*)

population dynamics   Populations are always under some form of environmental stress that threatens to disturb the stability of the population. The stress might be natural, such as extreme cold temperatures, or manmade, such as the impact of *pesticides.* Changes in the size of a population and the factors creating these changes are called population dynamics.

Populations respond to stress in four ways. The most obvious is an increase in the death rate (*mortality*) or a decrease in birthrate (*natality*). Organisms often move to other areas (*migration*) to avoid the stress. Organisms that reproduce rapidly and have many offspring might be capable of adapting to the stress through the process of *natural selection.* Any population that cannot successfully respond to the stress may perish from that *ecosystem* or even become *extinct.* (See *Population stability, Carrying capacity, Doubling-time in human populations*)

population ecology   Population ecology is the study of population statistics, the factors that affect a population, such as birth- and death rates, and the causes for changes in these rates. The size of a population in a specific area is called the population density and is central to understanding population ecology.

A population's density is controlled by two types of factors. Density-independent factors, as the name implies, have no correlation to the existing density. For example, extreme weather conditions can reduce populations, which has nothing to do with the population density. Density-dependent factors occur as a result of the existing population density. For example, the food supply will dwindle as a population grows, resulting in less food for more individuals. (See *Carrying capacity, Population dynamics, Population stability, Tolerance range, Limiting factors*)

population explosion   Refers to a dramatic increase in the size of a population, often referring to human populations. (See *Doubling-time in human populations*)

Population Reference Bureau (PRB)   The Population Reference Bureau (PRB) is a private scientific and educational organization that collects, interprets, and disseminates information on population trends. Established in 1929, the PRB serves as a link between scientific research and the public and those who affect policy worldwide. It offers many excellent publications and teaching materials. It has an information service and one of the oldest and largest "population" libraries in the world. Write to 1875 Connecticut Ave. NW, Suite 520, Washington DC 20009-5728.

population stability Populations within a community are always under some form of environmental stress. Natural stress factors include extremes in temperature, lack of moisture, or floods, while manmade stress includes the use of **pesticides,** and **deforestation.** Populations attempt to maintain a stable population. There are three aspects to this stability: 1) persistence is the ability of a population to resist change; 2) constancy is the ability of a population to maintain a certain size; 3) resilience is the ability of a population to bounce back to its original condition after being exposed to some form of stress. (See *Population dynamics*)

potable water Refers to **water** that is fit for human consumption. It must contain low levels of salts (**freshwater**) and have little or no animal wastes and bacterial contamination. Much of the drinking water comes from **groundwater aquifers.** The availability of potable water is a life-threatening problem in many regions of the world. One and a half billion people worldwide do not have regular access to potable water and about 5 million people die each year from waterborne diseases. (See *Drinking water, Water pollution*)

PP PP is an acronym for polypropylene, which is a **plastic** used in many diverse products, such as snack food packaging, disposable diapers, and the encasements for car batteries. It makes up about 10 percent of all the plastics produced in the U.S. and is rarely recycled. These products are stamped with the number "5" surrounded by a triangle made of arrows. (See *Plastic recycling, Plastic pollution*)

prairie See *Grassland.*

predator See *Predator-prey relationship.*

predator-prey relationship When one animal kills and eats another animal, the killer is called the **predator** and the animal killed is the **prey.** The interaction between the two is a predator-prey relationship. A bird eating a worm, a cat eating a mouse, and a lion eating a zebra are examples of predator-prey relationships. This type of interaction is an important aspect of **food webs**. (See *Symbiotic relationships, Energy pyramids*)

President's Council on Competitiveness The President's Council on Competitiveness was created to help American industry become more competitive. During 1991 and 1992, this council began to attack regulations designed to protect the environment and believed to be harmful to the economy. One facet of this attack was to issue new rules that reduced public access to information regarding environmental issues. For example, the **Clean Air Act** of 1990 originally required public notification and public hearings before any company could exceed approved levels of pollutant emissions. The Council on Competitiveness reviewed the act and removed the necessity of any public review before a company could increase levels of pollution.

Other environmental topics the council expressed concerns about include **wetlands** protection, disposing and storing of **hazardous wastes,** and testing and marketing genetically engineered products. (See *Economy vs. environment, Economic vs. sustainable development*)

prey   See *Predator-prey relationship.*

primitive technologies   Refers to aboriginal survival tools and techniques that demonstrate how humans have survived without the use of modernday high-tech equipment. These include weapons such as atlatls, rabbit sticks, bolas, and slings. More domesticated tools and techniques include the use of primitive hand drills, making fire from scratch, and the preparation of foods and medicines. These and other primitive technologies are taught at the Boulder Outdoor Survival School, Inc., in Flagstaff, Arizona.

producers   Organisms can be categorized according to how they obtain energy and nutrients to survive. Producers use radiant energy from the sun, during **photosynthesis,** to build complex organic molecules (sugars). These organic molecules contain the sun's captured energy in chemical form, which can be used by all forms of life. This chemical energy is passed along **food chains** by **consumers.** Without producers, all other life on earth would cease to exist. Trees, flowering plants, grasses, ferns, mosses, **algae,** and **phytoplankton** are all examples of producers. (See *Energy pyramids*)

profundal region   Refers to the level in a lake where sunlight does not penetrate. (See *Aquatic ecosystems, Standing water habitats*)

PS   PS is an acronym for polystyrene, which is a **plastic** used to manufacture foam paper cups and food containers. It makes up about 11 percent of the total plastics manufactured in the U.S. Some PS recycling programs are just getting underway. For recycling purposes, these products are stamped with the number "6" surrounded by a triangle formed by arrows. (See *Styropeanuts, Plastic recycling, Plastic pollution*)

pumice   Pumice is a porous volcanic rock that is mined for one reason—to give jeans a "stonewashed" look. Regions have been **strip-mined** with all the accompanying destruction to meet this need. Environmental groups have pressured jean companies to use alternative, less environmentally damaging methods to stonewash jeans. (See *Mineral exploitation*)

purse seine nets   Purse seine nets are fishing devices used to catch large numbers of tuna. In addition to the intended catch, the nets trap large quantities of fish and other marine life not harvested. This unintentional catch, called **bycatch,** is thrown back into the sea with most of the animals dead or dying.

During the 1950s, tuna fishermen discovered that yellowfin tuna are often found swimming below dolphin herds. (It's believed the tuna follow dolphins because of their superior ability to locate food.) Tuna boats routinely use a technique of encircling dolphin herds with purse seine nets and hauling in the tuna catch below, along with the herd of dolphins above. The dolphins usually die before being released. It's estimated that over 100,000 dolphins die in these nets each year.

Public pressure within the U.S. has resulted in many tuna canners selling "Dolphin Safe Tuna," which assures the consumer that the tuna was not caught in **driftnets** or purse seine nets. (See *Flipper Seal of Approval, Gillnets, Turtle excluder device*)

push-pull hypothesis   A theory stating that conditions within one country, such as poverty or unemployment, push people out (**emigration**), while conditions within other countries, such as a high standard of living and plenty of jobs, pull people into that country (**immigration**). (See *Immigration, replacement-level*)

PVC   PVC is an acronym for polyvinyl chloride, which is a **plastic** used for construction and in plumbing. It is the toughest of the commonly used plastics and not normally recycled. It makes up about 5 percent of all the plastics produced in the U.S. Incineration of this plastic produces toxic gases. (See *Plastic recycling, Plastic pollution*)

pyrethroids   Pyrethroids are one of the four categories of synthetic **insecticides.** Pyrethroids are synthetically produced versions of a **natural insecticide** called pyrethrin, which is found in chrysanthemums. They are considered a "soft pesticide" since they break down into harmless chemicals shortly after application, usually from a few days to a few weeks. Some examples of these synthetic pyrethrins are permethrin, which is sold under many brand names, and tralomethrin, sold under the brand name Scout. (See *Pesticides*)

qualitative studies Studies that concentrate on descriptions, as opposed to numerical analyses. (See *Ecological study methods*)

quantitative studies Studies that are based on counts, measurements, and other numerical analyses. (See *Ecological study methods*)

queenright Refers to a honeybee colony that contains a queen. A colony is dependent on a healthy queen since she alone can produce eggs. (See *Insects*)

quiescence Refers to a temporary resting stage (dormancy) that usually occurs due to unfavorable environmental factors, such as a lack of moisture for seeds to germinate or cold temperatures for **_insects_**.

R-strategists R-strategists are usually small organisms that are short-lived and produce large numbers of offspring and offer little or no parental care. These organisms expend a great deal of energy producing dozens, hundreds, are even thousands of eggs or live young. Their reproductive strategy is that there is safety in numbers. If enough offspring are produced, some of the young are likely to survive. Most insects and some reptiles, amphibians, and small mammals use this strategy.

   The population size of these organisms is controlled by density-independent *limiting factors,* which are factors that have little or nothing to do with the density of the existing population. Extremes in temperature, shortages of water, and unexpected population explosions of predators are examples of these limiting factors. The population size of R-strategists does not usually reach the environment's *carrying capacity* as it does with *K-strategists.*

R-value See *Thermal insulation, building.*

Rachel Carson Council, Inc. This council was founded according to *Rachel Carson*'s wishes to establish a nonprofit scientific and educational corporation that focuses on chemical contamination of our environment and its effects on human health. The council responds to inquiries from individuals and organizations around the world and provides information through publications, meetings, and government programs. Write to 8940 Jones Mill Road, Chevy Chase, MD 20815. (See *Pesticide dangers, DDT, Bioaccumulation, Biological magnification*)

radiation Some elements have unstable nuclei and are constantly throwing off particles, releasing the energy that once bound these particles together. The release of these particles produces radiation. There are three types of radiation. Alpha radiation consists of particles composed of two neutrons and two protons. These particles move rapidly, but only short distances and can be stopped by material as thin as paper. Alpha radiation is the most harmful form of radiation to organisms.

   Beta radiation consists of electrons, which also move short distances and can be stopped relatively easily. Gamma radiation is a form of **electromagnetic radiation,** like X rays, and travels greater distances and can pass through thick concrete walls. Electromagnetic radiation is caused by waves, as opposed to particles.

When something is struck by radiation, it is said to be irradiated. **Food irradiation** is currently used as a controversial method of preserving food. X rays for diagnostic purposes are another controlled use of irradiation. **Nuclear power** and nuclear weapons use radioactive elements such as uranium and plutonium. The radiation emitted by large quantities of these substances—while being used and after disposal—poses a significant threat to all forms of life on our planet. (See *Nuclear waste disposal, Radioactive waste disposal, Plutonium pits, Hanford Nuclear Reservation*)

radio–wave light bulb

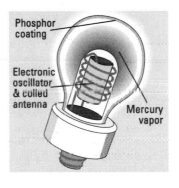

Phosphor coating

Electronic oscillator & coiled antenna

Mercury vapor

The radio-wave bulb (also called the E-lamp) uses a new technology to create light. It uses high-frequency radio signals instead of the usual filament to produce light. The bulb has a life expectancy of about 20,000 hours compared to an incandescent bulb's expectancy of 750 to 1,000 hours, and the new compact fluorescent light bulb, which is expected to last 7,000 to 10,000 hours. The E-lamp has an added advantage over compact fluorescent bulbs in that they fit into all existing sockets and can be used with a dimmer switch.

A 25-watt E-lamp bulb produces the same amount of light as a 100-watt incandescent bulb. The new bulb costs about 15 dollars, but the extended life translates to a substantial savings of about 20 cents per week over incandescent bulbs. Changing one incandescent bulb to an E-lamp would eliminate about one ton of carbon dioxide that would have been produced while generating the extra electricity. If these new light-bulb technologies are accepted and implemented, a major step will be taken to help reduce **greenhouse gas** emissions. (See *Bioluminescence*)

radioactive waste disposal

Even if totally safe **nuclear power plants** could be constructed, disposal of the leftover radioactive waste would still be a major problem. Up until 1983, most radioactive waste was dumped into the ocean. It is estimated that 90,000 metric tons of radioactive waste already lie on the ocean floor. Since the early 1980s, alternative methods of disposal have been suggested, studied, and tested, but none are in full operation.

Radioactive waste can be divided into high-level waste, which includes spent fuel rods from **nuclear power** plants and old nuclear weapons, and low-level waste, which includes any other radioactive materials from nuclear power plants such as the coolant, and from medical facilities that routinely use radiation for treatment and diagnostics.

Tons of high-level radioactive waste, including 21,000 metric tons of spent fuel rods, are stored in temporary facilities awaiting a final destination. The most accepted plan is to bury the waste deep in the earth's crust. An alternative to burial of high-level radioactive waste is to reprocess the material for re-use. Reprocessing is more expensive than storing, but eliminates the need for disposal. Reprocessing plants exist in France and the United Kingdom.

Low-level radioactive waste is usually placed into steel drums and shipped to special landfills. Some of these have been closed due to radioactive contamination of the groundwater. Much of the low-level radioactive waste is probably illegally disposed of.

Two permanent **nuclear waste disposal** repositories are scheduled to open, one for high-level wastes and the other for low-level wastes. Each state is required to establish low-level radioactive waste disposal sites. This edict from the federal government has caused a proliferation of the **NIMBY syndrome** throughout the country.

radon   Radon is a naturally occurring radioactive gas. It is odorless, tasteless, and invisible. It is released during the natural breakdown of uranium, which is found in many types of rocks and soils. When released into the air, it becomes diluted and harmless. If, however, the gas becomes trapped in an enclosure such as a basement, it accumulates and the concentration can become dangerous. The amount of radon that enters a building is based on the amount of radon present in that area and the construction of the building.

The problem was first discovered in the 1960s when high concentrations of radon were found in homes that were built with contaminated materials from uranium mines. This led to greater awareness, which revealed many contaminated homes across the country due to natural causes such as seepage into foundation cracks.

Exposure to radon is believed to increase an individual's risk of developing lung cancer. It is estimated that between 5,000 and

20,000 deaths from lung cancer may be blamed on radon exposure. People exposed to roughly 10 times the normal indoor concentration of radon gas have about the same likelihood of developing lung cancer as someone who smokes half a pack of cigarettes a day. Long-term exposure to low levels of radon is thought to be more dangerous than short-term exposure to high levels of the gas.

Although only a small percentage of buildings have the problem, a simple test can determine if danger exists. Towns and local agencies often provide test kits, or professional services can be contracted. Homes found to be contaminated can usually have the problem corrected by structural changes. (See *Indoor pollution, Radiation, Healthy homes*)

radura The radura is the government-approved symbol required on all foods that have been irradiated. The symbol was originally created in the Netherlands back in the 1970s. It bears a striking resemblance to the Environmental Protection Agency's logo. It appears as a broken circle that contains a highly stylized plant. (See *Food irradiation, Radiation, Nuclear waste disposal*)

Rails-to-Trails Conservancy The Rails-to-Trails Conservancy converts abandoned railroad tracks into trails used for **bicycles,** hikers, and naturalists to enjoy. Many of these trails (also called **greenways**) connect metropolitan areas, rural communities, and local parks. Some offer the commuter alternatives to congested highways.

With more than 3,000 miles of railroad track abandoned each year, there is the potential to offer viable commuting alternatives to automobiles. The Rails-to-Trails Conservancy is currently working with 350 municipalities to acquire and convert these tracks. There are already 3,187 miles of these trails operating across the country. Write to 1400 Sixteenth St. NW, Suite 300, Washington DC 20036.

rain forest destruction *Tropical rain forests* are some of the richest, most productive **ecosystems** on the planet, playing vital roles in **biogeochemical cycles** and containing the majority of the world's **biodiversity.** While covering only 7 percent of the earth's surface, they are home to over half of all forms of life on the planet. Rain forests are being destroyed at a staggering rate. It is believed that about 80,000 square kilometers (about the size of Austria) is converted to nonforest use every year.

These forests are often destroyed during **slash-and-burn cultivation** to make room for crops. The vegetation is burned to add nutrients to the soil for crops. This type of ***organic fertilizer*** enriches the soil for only a few years before becoming useless. Many forests in Central America have been cleared to make room for pastures to raise cattle for beef. Forests are also destroyed by ***clear-cutting*** for timber.

The effects of this destruction are numerous, including the impact on ***global warming, soil erosion*** and the loss of biodiversity. (See *Deforestation, Energy pyramids*)

range   The geographical distribution of a species is called its range.

realms   Refers to a method of describing animal distribution within major regions of the planet. (See *Zones of life*)

recycling   Recycling is the re-use of materials that have been recovered from the ***waste stream.*** This may be as simple as leaving your lawn mower clippings on the ground instead of being bagged (called ***grass cycling***) or as complex as recycling some plastics, such as ***styropeanuts,*** into new products such as wastebaskets. Recycling reduces the volume of ***municipal solid waste,*** which means there is less to put into ***landfills*** or ***incineration*** plants.

In addition to reducing what must be disposed of, recycling also reduces the need for ***mineral exploitation*** and the depletion of all our ***natural resources.*** Numerous statistics are available, but a few examples can paint the picture. The U.S. imports almost all of the aluminum it uses in manufacturing, but each year it throws away aluminum worth about $400 million, most of which could be recycled and be back in another aluminum can within six weeks. One Sunday issue of the *New York Times* uses between 60,000 and 75,000 trees, but the U.S. recycles less than 20 percent of all its paper waste.

There are many aspects to a successful recycling program. The logistics involve establishing methods of collection, separation, and transportation. Technology must be available to process the used materials, prepare it for re-use, and then remanufacture it into new products. Markets for the new product must be identified and available. Collecting and processing accomplish nothing if these raw materials don't have a market. Finally, incentives, enforcement, and educational policies must be established to complete the process.

Recycling programs have been initiated with everything from appliances to computer printer toner cartridges. Some examples are as follows: Corrugated cardboard boxes are recycled into cereal and shoe boxes. This week's newspaper is recycled into next month's newspaper. Computer printout paper is recycled into toilet paper. Glass food and beverage containers are recycled into new bottles plus glass wool and highway reflectors. Metal food containers are turned into new cans and household appliance parts. *PET* plastics can be turned into fiberfill for jackets, carpeting, and other new plastics. *HDPE* plastics can become auto bumpers and drainage pipes, and scrap metal can become new car parts and refrigerators.

Recycling programs have been mandated in many cities and states. Even with these programs and a new public awareness, only about 11 percent of the solid waste in the U.S. is recycled, compared with countries like Japan, which recycles almost half of its solid waste. (See *Plastic recycling, Appliance recycling, Copier toner cartridges recycling, Tires, recycled*)

recycling logos Three twisting arrows chasing one another are often marked on packaging by manufacturers to symbolize the product has been recycled or is recyclable. Three arrows indicates the product is supposedly recyclable. Three arrows within a circle supposedly mean the products have been recycled. The logo was originally designed by a Dutch paper company in 1973 but has become the universal symbol of recycling. These logos, however, are not regulated and relatively worthless as purchasing guidelines. (See *Paper recycling, Seals of approval, environmental; Green marketing, Green products*)

Paper product is made from recycled paper

Paper product is recyclable

red tides Refers to a sudden and dramatic population explosion of microscopic algae (*phytoplankton*) along the coast and often in harbors. This may be caused naturally but is usually attributed to human activities that produce imbalances in nutrient cycles. Raw or partially treated *sewage* dumped into the water or *fertilizer* carried by rainwater runoff adds excessive nutrients, causing the increase in the plankton.

Although called red, they may actually be brown, yellow, or green. They are all harmless organisms. However, the slimy, thick mats that form can block the sunlight from reaching the waters below, affecting the *aquatic ecosystem.* When these masses of algae die, the bacteria that feed on the decomposing algae have a population explosion of their own, reducing the amount of oxygen in the water. This kills large numbers of fish and other organisms. In some instances the algae produces toxins that poison shellfish and, in turn, can poison those eating the shellfish. (See *Estuary and coastal wetlands destruction, Eutrophication, Water pollution*)

reforestation   Refers to replanting stands of trees in areas that have been logged. The amount of time required between harvests of these trees is called the logging cycle. Most logging cycles today are about 60 years but could be as long as 150 years. Reforestation assures a continuous source of timber.
   Ancient forests that contain trees hundreds or even thousands of years old, of course, can never be replaced. The **biodiversity** of a forest is also believed to never return to its original richness once a virgin forest has been logged. (See *Deforestation, Forest Service*)

refrigerator doors   Conserve energy by checking the seal on your refrigerator door. Close the door on a piece of paper that is halfway in and half out of the refrigerator. If you can easily pull the paper out, you're wasting energy and your money. The seal may need to be replaced or the door adjusted. (See *Thermal insulation, Super-insulation, Tub bath vs. shower, Domestic water conservation*)

REM (Roentgen Equivalent Man)   Exposure to **radiation** is measured in units called REMs (Roentgen Equivalent Man). A typical X ray of the intestines releases 1 REM. Exposure to low levels of radiation (less than 10 REMs/yr) is believed to be safe but remains debatable.
   Moderate levels of radiation (10 to 1000 REMs/year) are believed to increased the likelihood of disease such as some cancers. High dosages (above 1000 REMs/year) will cause ill effects and the likelihood of death. (See *Nuclear reactor safety and problems, Food irradiation*)

remediation of hazardous waste   Refers to the process and business of cleaning up **hazardous waste** sites. Remediation may be as simple as manually mopping up an area or involve high-tech equipment and sophisticated chemical processes. New methods of remediation include the use of microbes, called **bioremediation,** that eat hazardous wastes. **Phytoremedia-**

tion uses plants to absorb hazardous wastes such as *heavy metals* or *radiation.* Reports show hazardous waste remediation to be one of the fastest-growing sectors of the U.S. economy. (See *Hyperaccumulators*)

renewable energy Renewable energy sources are considered to be inexhaustible, even if continuously used by man. Renewable energy and *nuclear power* are the two major alternatives to *fossil fuels.*

Renewable energy includes: *solar power, wind power, hydroelectric, geothermal energy,* and *biomass energy* sources. Renewable sources are usually nonpolluting and produce no hazardous materials as does *nuclear power.* Renewable energy can produce energy in the form of electricity, heat, and transportation fuel as opposed to nuclear, which produces only electric power. Many renewable sources are already cost-competitive compared to fossil fuels and will become even less expensive when used on a larger scale.

A problem facing many renewables is that of energy storage. Most of these sources are variable in nature. Solar availability varies due to day and night cycles and short-term weather conditions; wind due to short-term weather conditions; and hydro due to long-term weather conditions such as drought. Energy storage technologies are improving rapidly, and many of the inadequacies should be resolved within the next few years. Many existing renewable power plants use fossil fuels for backup purposes during downtime and peak usage periods.

Renewable energy sources supply over half of Japan and Israel's energy needs, while in the U.S., all forms of renewable energy fulfill only 7.5 percent of our needs. The *Environmental Protection Agency (EPA),* in a 1989 study, suggested that by the year 2050, renewable forms of energy can and should supply 30 to 45 percent of our energy needs, while many respected scientists believe the percentage can be much higher. However, the Department of Energy released a study in 1989, projecting that only 12 percent of our energy will be supplied by renewables in the year 2010. The amount of funding available for research and pilot programs over the next few years will dictate the actual reliance on renewable energy sources in the future. Unfortunately, the amount of research funding for alternate energy dropped from about one billion dollars in 1981 to less than one tenth that in 1990. (See *Cool energy*)

resilience, population See *Population stability.*

resistant crops   One alternative to **pesticides** is the use of genetically engineered crops that resist pests. Characteristics that naturally protect the plants from pests are selectively engineered or "bred" into the plant. This may include such modifications as shortening the plant's growth cycle to avoid a late season pest, or changing the genetic makeup to make plants resist pests they normally could not. For example, wheat rust is no longer a problem since strains resistant to this fungus have been developed. Resistant crops are often used as an important part of an **integrated pest management** plan.

Resource Conservation and Recovery Act of 1976   An important piece of legislation that defines how **hazardous wastes** can be transported, stored, treated, and disposed. It empowered the **EPA** to protect the country's water and air from contamination. (See *Superfund*)

respirable suspended particulates (RSP)   RSPs are airborne particles small enough to be inhaled deep into the lungs, but large enough to get stuck and remain there indefinitely. Since the nose, throat, and bronchial tubes filter out particles larger than 1.5 microns (a micron is one-millionth of a meter), and particles less than .1 micron are usually exhaled normally, RSPs range in size between .1 and 1.5 microns. **Asbestos** and **lead** are two examples of RSPs. Cigarette smoke particles carrying benzo-(a)-pyrene (a component of the incomplete combustion of tar) is an RSP and probably due in part for the increased risk of lung cancer among cigarette smokers. (See *Indoor pollution, Environmental tobacco smoke, Air pollution*)

respiration   Green plants perform **photosynthesis** to build sugar. These sugar molecules act as fuel for both plants and animals. Respiration is the process in which the molecules of sugar are burned (broken down) to release the energy stored within. This chemical energy is used by the organisms to survive (grow, move, reproduce, etc.). Respiration occurs when the sugar (glucose) molecule combines with oxygen, resulting in the formation of carbon dioxide, water, and the release of energy. (See *Carbon cycle, Food chain*)

ribbon sprawl   Refers to a type of **urban sprawl** in which commercial and industrial growth occurs along transportation routes such as major roads or commuter rail lines.

Richter scale A scale used to measure the intensity of geological disturbances ranging from 1.5 for barely detectable tremors to 8.5 for catastrophic earthquakes. The scale is logarithmic. A reading of 7 is 30 times greater than a reading of 6.

rill erosion Refers to soil erosion caused by rain and overflowing surface waters, producing a series of grooves or channels.

Rio Declaration on Environment and Development This document is one of the five documents debated at the ***Earth Summit*** in Rio de Janeiro in June 1992. This declaration was the result of a compromise between the industrialized countries and the "Group of 77" developing countries. The developed countries simply wanted to accept the principles put forth in a previous declaration called the Stockholm Declaration, written in 1972, stating the need to protect the planet. The developing countries, however, lobbied for more detail regarding specific concerns. They wanted the Rio Declaration to declare the need for financial and technical assistance to developing countries and to acknowledge that the industrialized countries were mostly responsible for the environmental problems that currently exist.

This document was not signed and therefore not legally binding. However, it is hoped that world governments will adopt the 27 principles put forth in the Rio Declaration.

risk assessment, environmental Refers to the process of gathering facts and making assumptions about the potential dangers to human health or to the environment caused by some project, product, or technology. These dangers could include various forms of ***pollution, radiation, pesticide*** poisoning, and effects on ***ecosystems*** or ***biogeochemical cycles.***

Risk assessment attempts to differentiate between true and perceived risks. Proper risk assessment should lead to ***risk communication*** to educate the public, and ***risk management,*** in which the knowledge gained is used to eliminate or reduce dangers, look for alternatives, and balance the economic advantages with the risks. (See *Risks, true vs. perceived*)

risk communication, environmental Risk communication is the process of providing the public with an understanding of any environmental risk associated with the new project, product, or technology. Public hearings or other forums give the public an opportunity to voice their opinions and influence the final decisions. (See *Risk assessment, environmental; Risk management, environmental; Risks, true vs. perceived*)

risk management, environmental   Environmental risk management studies the balance between the advantages and disadvantages of a new project, product, or technology. Managing environmental risk means weighing negative environmental impact against economic gains, and includes public interest and legal issues. Risk management is difficult since much of the information gleaned from studies cannot be based on fact, but by assumption. In some cases, what's good for the economy may be harmful to the environment, and the correct decision is highly subjective. (See *Risk assessment, environmental; Risks, true vs. perceived*)

risks, true vs. perceived   The public often has great difficulty determining the true risks of a new environmental project, product, or technology when they are bombarded with information—some factual, some not. This information comes from industry, government, or environmental groups. An informed public is the best way to assure that *risk management* is based on the best interests of the citizens. The difficulty involved in differentiating true versus perceived risks is underscored by comparing what the EPA and the public believe to be an environmental risk.

For example, the EPA believes that *indoor air pollution,* unsafe *drinking water,* and workers' exposure to on-the-job chemicals rate as "very high" risks to our health, but the public considers all three to be "low" risks. The public, however, thinks chemical manufacturing plants are a "very high" risk, but the EPA considers this category only a "mid-level" risk. Both do agree, however, on the dangers of outdoor *air pollution* (high) and accidental *oil spills* (medium).

RIYBY   Acronym for Recycling In Your Backyard, which describes a new movement to make *recycling* a small business opportunity instead of just a large corporate or municipal enterprise. There are many books and magazines about small recycling business opportunities. (See *In Business, Garbage*)

rodenticide   Refers to a *pesticide* that kills rodents such as mice and rats. The most common rodenticide is warfarin, which, when eaten by animals, causes internal bleeding and death. Warfarin is not selective, meaning it will kill or make seriously ill animals other than the targeted pests. (See *Nontarget organism*)

rookery Refers to a breeding or nesting site for animals, usually birds.

room-and-pillar method Refers to a method used to provide structural support in **subsurface mining.**

RTC (Resolution Trust Corporation) The RTC is responsible for selling off 45,000 properties that have been taken over by bankrupt savings and loan institutions. It is estimated that 3 to 5 percent of these properties are undeveloped lands that are inhabited by **endangered species** or rare natural **habitats** worth saving. **Environmental organizations** are trying to identify these properties and in many cases procure them for protection or recreation. For example, the conservation group the Trust for Public Land (TPL) purchased 100 acres of mountainside property from the RTC and turned it over to the city of Tucson for use as a public park.

running water habitats **Freshwater ecosystems** can be divided into those found in **standing water habitats** and those found in running waters, which include streams and rivers. Since the water is always moving in running waters, organisms attached to rocks and other substrate play the most important role instead of floating **plankton,** which are important in other **aquatic ecosystems.**

Most moving bodies of water are usually shallow, with sunlight penetrating to the bottom, allowing plants to grow and act as food for organisms such as insects; this establishes a grazing **food chain.** Predatory insects and fish feed on these insects.

Nutrients are also added to the stream by dead and decaying organisms (or parts of organisms) falling into the water. Leaves, twigs, and bark from trees fall in, and animal waste and other organic particles are washed in. This organic matter is attacked by bacteria, fungi, and microorganisms, which establish a decomposing food chain.

Insects join in the attack, mainly to eat the bacteria and fungi that are attached to the decaying leaves. The decaying leaves and associated organisms flow downstream and eventually sink to the bottom to become a food source for many other organisms, especially aquatic insect larvae. (See *Running water habitats, human impact on*)

running water habitats, human impact on
Freshwater ecosystems can be divided into ***standing water habitats*** and running water habitats. Running water habitats, which include rivers and streams, have been used by humans for hundreds of years as toilet bowls, where pollutants and waste could be "flushed" out of sight. The vast majority of rivers in the U.S. are polluted to some degree. Although streams and rivers can cleanse themselves when low levels of waste are introduced, it is the volume and speed with which they are introduced that cause irreparable harm.

   Industrial water pollution pours huge quantities of ***toxic waste*** and by-products into rivers. Water is drawn from rivers to cool power plants, which is then returned at temperatures that destroy ecosystems, called ***thermal water pollution.*** Raw ***sewage*** is poured into streams in vast quantities in short periods of time, changing the chemical makeup of the water and often eliminating the usual inhabitants. ***Pesticides*** and ***fertilizers*** are washed into the water, changing the chemical composition of the water and killing aquatic organisms. Radioactive waste has also been found in the sediment of streams and rivers. (See *Radioactive waste disposal, Water pollution*)

runoff   That portion of precipitation that is not absorbed by soil and washes away is called runoff. Runoff collects in streams and rivers, then finally makes its way to the ocean. Storm-drain runoff can contain numerous kinds of pollutants, including gasoline, ***de-icing road*** salts, ***plastic*** products, and just about anything you see in a street. Agricultural runoff refers to water that washes off of irrigated land and usually contains large amounts of ***fertilizers*** and ***pesticides. Acid mine drainage*** runoff comes from mining operations and carries dangerous acids into bodies of water. (See *Pointless pollution, Water cycle*)

S-curve If a **population** was allowed to grow unchecked, it could take over the entire planet. If you were to plot a graph of this unchecked growth, it would look something like the letter "J" and is therefore called a J-curve. Resources such as food, however, are limited, and conditions such as temperature and moisture are never ideal. Most populations, therefore, usually start by looking like a J-curve, but then reach the maximum size that the environment can support, called the **carrying capacity,** and then level off. This creates a curve that looks something like the letter "S," hence the S-curve of population growth. (See *J-curve, Limiting factor*)

Safe Drinking Water Act Created in 1974, this law required the **EPA** to establish maximum contaminant levels for all drinking water pollutants so that they do not pose a danger. Although it has reduced the amount of microbiological contaminants, it has done little to protect us from water contaminated with **pesticides** and toxic substances. The law has been modified many times and was reauthorized in 1992. (See *Water pollution, Toxic waste disposal, Aquifers*)

Saint Elmo's fire Refers to a glow produced by discharging atmospheric electricity. It usually appears at the point of structures such as church steeples, airplane wings, and boat masts.

salinity Refers to the concentration of dissolved salts in water, usually measured in parts per thousand, ppt. (See *Aquatic ecosystems*)

salinization, soil Salinization of soil is a harmful side effect of **irrigation.** When irrigation water washes over the surface of the soil, it dissolves and collects salts from the earth, making the water salty. When the excess water evaporates, it leaves behind increased concentrations of these salts in the soil. This process is called soil salinization.

In arid regions (where most irrigation occurs), evaporation occurs quickly and soil salinization progresses rapidly, stunting the growth of crops and eventually making the land useless. About one-fourth of the world's crops are believed to be growing at reduced levels because of soil salinization.

Methods of renewing salinized soil include flushing the region with excessive amounts of water to remove the salts, but this often results in excessive concentrations of salt in *aquifers* or *water-logging.*

saltatorial   Refers to organisms that move by hopping, leaping, or bounding. For example, grasshoppers are saltatorial.

saltwater intrusion   When *groundwater* is removed faster than it can be replaced, the water table drops. When this occurs in coastal areas, *aquifers* may become recharged (refilled) with salt water from the ocean instead of fresh water. When salt water replaces fresh water in aquifers, it is called saltwater intrusion. Saltwater intrusion has become a problem in many coastal areas such as California, New York, and Florida. (See *Water mining*)

sand   Refers to particles of *sediment* that do not adhere to each other and are between .0625 and 2.0 millimeters in size. Sand types are divided into very coarse, coarse, medium, fine, and very fine. (See *Soil types*)

saprophyte   Refers to a plant that obtains nutrients from dead, decaying organic matter, such as many *fungi* (mushrooms). (See *Food webs*)

savanna   Savannas are one of several kinds of *biomes.* The primary factors that differentiate biomes are temperature and precipitation. Savannas (also called tropical grasslands) are similar to *grasslands* in some respects but receive more precipitation—between 75 and 150 centimeters (30 to 60 inches) per year in a seasonal pattern.

The environment appears similar to that of grasslands but contains some trees. Since fires often destroy these regions, most of these trees are fire resistant. Some of these trees are capable of *nitrogen fixation.* The trees provide a habitat for animals that otherwise could not survive in the region. Most animals are *grazing* mammals or rodents, birds, reptiles, and insects. The larger mammals differ depending upon the location of the savanna. Australia has kangaroos, Africa has antelope, and South America has llamas.

Save the Dunes Council This council was founded in 1952 to preserve and protect the Indiana Dunes for public use and enjoyment. The Indiana Dunes were formed over the past 15,000 years by glaciers, water, and wind. Walking through these dunes, one can see how the communities of plants and animals have changed as the waters of an ancient glacial lake receded to form what is now Lake Michigan. Research on the dunes continues to contribute to our knowledge of ecology.

The council was instrumental in protecting the dunes from development. In 1966, Congress created the Indiana Dunes National Lakeshore as a unit of the National Park System. Presently, the council is working toward expanding the park boundaries and continuing to protect it from future development. Write to 444 Barker Road, Michigan City, IN 46360.

Save-the-Redwoods League The Save-the-Redwoods League was founded in 1918. Its goal is to acquire and protect **Coast Redwood** and Giant Sequoia forests. Since the boundaries of many existing Redwood parks are inadequate to maintain the Redwood **ecosystem,** the **watershed** lands surrounding these parks must also be acquired. Write to 114 Sansome Street, Room 605, San Francisco, CA 94104. (See *Ancient forests*)

scatology The study of animal feces. (See *Scavengers, Decomposers*)

scavenger These are animals that obtain nourishment by eating organisms that have recently died. Scavengers are usually the first to arrive at an animal carcass and begin the process of decomposition, which is then taken over by the **decomposers.** (See *Food webs*)

seals of approval, environmental The public's interest in environmental issues has resulted in many claims about how friendly products are to the environment, both in how they are produced and how they can be discarded. Most of these **green marketing** claims are gimmicks based on marketing research instead of scientific research.

Many organizations and companies are trying to formalize and

S.C.S.
Green Cross
Certified

GREEN SEAL

establish meaningful seals of approval to assist the consumer. In the U.S., two new organizations have emerged: Green Cross and Green Seal. The Green Cross uses a fat green cross over the corner of a blue globe, and Green Seal uses a long green check mark across a blue sphere. Both of these companies have their own set of environmental standards and, for a fee, will examine manufacturing plants, test a company's products, and then certify it by affixing their seal to the product.

The federal government is also getting involved with the Environmental Marketing Claims Act, which will give the EPA its own green marketing guidelines and the ability to enforce when companies can and cannot use words like "recyclable" and "biodegradable."

Other countries are also active in standardizing environmental claims. Germany has the Blue Angel seal, Canada has the Environmental Choice seal, and Japan has the EcoMark seal. (See *Flipper Seal of Approval, Green products*)

secondary air pollutants   Burning fossil fuels to drive cars and to generate electricity produces five primary air pollutants responsible for **air pollution. Secondary air pollutants** are substances created when the primary pollutants combine with one another or some other substance. This reaction is driven by energy from the sun. For example, sulfur dioxide, a primary pollutant, reacts with oxygen and moisture to produce sulfuric acid, a secondary pollutant.

Photochemical smog is produced when nitrogen compounds (a primary pollutant) produce ground-level **ozone** (a secondary pollutant). (See *Acid rain*)

sediment   Refers to **soil particles** that have been transported, usually by water.

selective harvesting   Refers to a logging method in which only certain types of trees are removed from a forest. For example, a certain species of tree or only mature trees are harvested. Selective harvesting causes much less harm to a forest ecosystem than does **clear-cutting.** (See *Forest management, integrated; Forest Service; Reforestation*)

septic tank   An underground tank designed to accept, hold, and decompose the contents of domestic wastewater. Bacteria break down the waste, leaving an organic *sludge* that settles to the bottom. The *effluent* then flows out of the tank into the surrounding ground. About 24 percent of U.S. homes use septic tanks. (See *Domestic water use, Water treatment*)

sere community   The *communities* of organisms found in an area gradually change over time, beginning with a *pioneer community* and ending with a climax community. This process is called *succession.* Each community that forms during this transition is called a sere community, and the entire sequence is called a sere. (See *Succession*)

sewage   Refers to wastewater flowing through a sewer. If the waste originates from *domestic water use* (residences or office buildings), it is called domestic sewage. If it is from *industrial water use* such as manufacturing facilities, it is called industrial sewage. If it is from storm drainage systems, it is called storm sewage. Sewage pipes carry the waste to *sewage treatment* plants or directly into a stream, river, or larger body of water, causing *water pollution.*

sewage treatment   Refers to treating all forms of wastewater that have been collected as *sewage.* The treatment renders wastewater less harmful before it is released into the environment. (Some sewage is released directly into the environment and is therefore never treated.)

Sewage treatment is usually divided into three stages: primary, secondary, and tertiary sewage treatment. Treated water in the U.S. must use primary and secondary sewage treatment processes. Since primary and secondary don't remove large amounts of some substances such as phosphates, nitrates, *heavy metals,* or *hard pesticides,* a few municipalities also use a tertiary sewage treatment process.

Primary treatment filters and settles out large particles, including everything from sticks to sand, but leaves the organic material in the water. Bacteria and other microbes begin to multiply and decompose the organic matter.

Secondary treatment holds the treated water until the microbes can complete their job of decomposing the organic matter. Secondary treatment encourages microbe growth by aeration since these microbes require oxygen. The microbes use the organic matter as food and incorporate it into their own bodies. Since the microbes are larger than the organic matter, they settle out of the water and are removed. This mass of microbes (both alive and dead) and any remaining waste are called **sludge.**

If the water is to be released following the secondary treatment, it is first disinfected to reduce the number of remaining microbes still in the water. This is usually accomplished with **chlorination.** Tertiary water treatment might be used by some municipalities and industries to reduce remaining pollutants and contaminants. Some industries must use tertiary treatments to remove contaminants specific to their industry. (See *Industrial water pollution, Waste-to-energy power plants, Composting, large-scale*)

sex attractants   Some organisms produce **pheromones,** which are substances produced by individuals of a species to communicate information to other members of the same species. Some of these pheromones act as sex attractants, used by the females to attract males. Most studies on sex attractants have been performed on insects and have unique applications.

An interesting alternative to using **pesticides** is the use of synthetically produced sex attractants (pheromones) that mimic the real thing. These artificial sex attractants are sprayed in the infested area. The abundance of the attractant confuses the males and drives them "crazy," searching for the females. Since the males cannot find the females, young are not produced and the pest population dwindles in size. Sex attractants are now available for about 30 insect pests. Some of these attractants can be detected by males almost two miles away.

A common residential scene in some areas of the country includes yellow gypsy moth traps, which contain a sex attractant under the brand name Gyplure. Male moths are attracted to the lure and become trapped. Female gypsy moths cannot fly, so they are doomed to remain outside the trap, while the males are stuck inside. Sex attractants are most effective when used as part of a comprehensive **integrated pest management** plan. (See *Insect sterilization*)

sex ratio   A population's birthrate and death rate are dependent, in part, on two factors: **age distribution** and sex ratio. The sex ratio is the number of males relative to the number of females. With humans, about 106 males are born for every 100 females. The ratio differs wildly among different types of organisms. There are many insects, for example, whose **populations** are almost entirely female, as with aphids.

sexual reproduction   When new individuals are created by uniting two cells, called gametes, it is called sexual reproduction. In most plants and animals, the two cells come from two individual parents, but in some cases, they are both by the same individual. The gametes may unite by internal fertilization, as with humans, or come in contact with each other externally. External fertilization can occur via the water (as with many fish) or the wind (as with many grasses). Most flowering plants must rely on insects to transport these cells for them (pollination). (See *Asexual reproduction, Parthenogenesis*)

sheet erosion   Pertains to the erosion of a uniform layer of surface soil from a large area caused by **runoff** water. (See *Soil conservation, Monocultures*)

sick-building syndrome (SBS)   Sick-building syndrome refers to a collection of symptoms people get while in tightly closed buildings with poor ventilation due to inefficient **heating, ventilation, and air-conditioning (HVAC)** systems. These symptoms cannot usually be traced to any one cause, as with **building-related illness,** but are linked to a building's overall environment. The symptoms are usually similar to a cold or flu, including dry or burning sinuses and runny eyes and noses. Often large numbers of people are affected with the same symptoms at the same time within the building. The symptoms are probably caused by low levels of a variety of contaminants in the building. The number of SBS complaints has been soaring in recent years. (See *Baubiologie, Indoor pollution*)

Sierra Club   The Sierra Club was founded by **John Muir** and recently celebrated its 100th birthday (in 1992). Its first goal was to help preserve the beauty of the Sierra Nevada mountain range. Since that time, the Sierra Club has grown to 650,000 members and has played a significant role in the formation of America's **National Park and Wilderness Preservation Systems,** protecting over 132 million acres of public land. It concentrates its preservation efforts on clean air and water, safe disposal of toxic wastes, and energy and population issues. It files lawsuits and lobbies agencies to further its goals. It has helped pass over 100 pieces of legislation over the past 10 years, and helped to pass **Superfund** legislation for **toxic waste** cleanup. It has numerous publications and is considered the world's largest conservation publisher. Write to 730 Polk Street, San Francisco, CA 94109.

silent killer, the   See *Asbestos.*

silviculture   The practice of growing and tending forests.

sink holes   When **groundwater** is substantially reduced due to **water mining** or drought, the once-saturated earth can collapse, forming a large depression in the land's surface called a sinkhole. This usually occurs in regions where **karst terrains** are common. Some sink holes can be large enough to destroy entire homes. (See *Aquifers*)

SLAPP   An acronym for Strategic Lawsuit Against Public Participation. The First Amendment of the U.S. Constitution guarantees an individual's right to participate in any peaceful legal action to influence his or her government. This includes actions such as signing petitions, writing letters, reporting violations of the law, public demonstrations, public hearings, boycotts, and the like. A SLAPP is a lawsuit filed by someone with opposing views, whose sole purpose is to stifle these activities—in other words, make it very uncomfortable and expensive to speak out in opposition of (or in favor of) some issue.

   Hundreds of such suits have been filed in efforts to stifle environmentalists expressing their opinions about environmental issues. People who have protested against developers, utilities, and private companies have been hit with SLAPPs. More recently, SLAPPs have been used against environmental organizations such as the Nature Conservancy and the Sierra Club. (See *SLAPP BACK*)

SLAPP BACK   In response to the growing threat of **SLAPP**s, many individuals and environmental organizations have successfully filed lawsuits "against" those who filed the original SLAPP, now called a SLAPP BACK. For information about SLAPPs and SLAPP BACKs contact: Coalition Against Malicious Lawsuits, P.O. Box 751, Valley Stream, NY 11582; SLAPP Resource Center, University of Denver, College of Law, 1900 Olive Street, Denver, CO 80220; Citizens' Clearinghouse for Hazardous Waste, P.O. Box 6806, Falls Church, VA 22040; First Amendment Project/California Anti-SLAPP Project, 1611 Telegraph Avenue, #1200, Oakland, CA 94612.

slash-and-burn cultivation   People in many less developed countries use an old practice called slash-and-burn cultivation to grow gardens and crops. A small area within a forest is cleared of vegetation and then burned. Since forest soils are naturally poor in nutrients (the nutrients are all up in the trees), the dead, burned vegetation acts as an *organic fertilizer,* enriching the soil so crops can grow. The cleared land, however, can be used for only a few years before the nutrients are depleted and the area abandoned. As long as small tracts are cleared, they are replaced with new growth, but larger expanses of forests cleared in this manner result in *soil erosion* and severe damage to the *forest ecosystem.* (See *Rain forest destruction, Deforestation*)

sludge   Sludge is the thick, gooey mass of microbes, organic matter, and other solids that is removed from wastewater in *sewage treatment plants* before the water is released into the environment. The volume of sludge produced is staggering. The San Francisco Bay area produces about 2,500 metric tons of sludge each day. (See *Sludge disposal, Biomass energy*)

sludge disposal   Once *sludge* has been collected from sewage treatment plants, it must be disposed of. Beginning in the 1920s and continuing through the late 1980s, the ocean was the sludge dumping ground. Millions of tons of sludge have been dumped within a few miles of shore, destroying the fish and shellfish industry in many areas. Almost all sludge is now disposed of on land, either in *landfills, incineration* plants, or in a few instances turned into *compost.* Some municipalities now pretreat sludge to remove *toxic waste* substances such as *heavy metals* and use it as a building material to produce brick, cardboard, and paving material. (See *Biomass energy, Waste-to-energy power plants*)

Smithsonian Institution   Most people know of the Smithsonian in Washington DC for its wonderful museums and the National Zoo. In addition to educating people of all ages, the institution is also involved in conservation and environmental research. These studies are conducted in research facilities around the country and many parts of the world.

The institution uses over 6,000 volunteers each year to teach at and staff museums and the zoo. It also publishes educational materials for everyone from the scientist to the layperson, as well as the colorful *Smithsonian* magazine. The institute was founded in 1846 and has 2.6 million members. The annual membership fee is 20 dollars. Write to The Smithsonian Institution, 1000 Jefferson Dr. SW, Washington, DC 20560. (See *Environmental education*)

smog   Smog originally referred to the combination of smoke (from smoke stacks) and fog. Today, however, it refers to any visible accumulation of substances that causes **air pollution.** Pollutants emitted directly into the air are called primary pollutants, and others, produced by chemical reactions under the influence of the sun, are called **secondary air pollutants.** (See *Photochemical smog, Industrial smog, Dust dome, Urban heat island*)

smoke detectors   Some smoke detectors are called "ionization" detectors because they use minute quantities of americium-241, which is a radioactive by-product from processing plutonium for nuclear weapons. As long as the seal within the device is not broken, the **radiation** will not be released. The problem, however, is what happens to all the americium when these smoke detectors are disposed of. **Landfills** containing this radioactive substance, which is dangerous for hundreds of years, could leak into the soil and **groundwater** supply. There are alternative types of smoke detectors that do not use radioactive americium.

snags   Old, dead standing trees in a forest. (See *Forest management, integrated; Forest Service*)

sociobiology   The study of the biology of social behavior.

soft pesticides   Refers to **pesticides** that decompose into harmless chemicals within a few hours or days. (See *Organophosphates, Carbamates*)

soil   Soil is a mixture of minerals (inorganic matter) and dead, decaying organisms (organic matter) that forms a thin layer over the earth's surface. Soil also contains air, moisture, and countless **soil organisms.** Soil is responsible for nourishing and supporting plant growth. It is composed of a mixture of different types of **soil particles.** About 14,000 **soil types** have been classified in the U.S.

soil conservation   One-third of the topsoil originally found on croplands in the U.S. is gone due to **soil erosion** by water and wind. Soil washes away and wind blows off topsoil from our croplands seven times faster than soil is created naturally. Soil conservation refers to the techniques and practices that help stabilize soil to reduce this erosion.

Soil conservation methods include the following: **conserva-**

tion tillage, contour farming, strip cropping, terracing, alley cropping, gully reclamation, windbreaks, and *zoning.* These conservation methods are used on about one-half of all U.S. croplands.

Almost 50 percent of all human-caused soil erosion, however, is not due to agriculture. Much of it comes from other factors such as *clear-cutting* of forests, *urban sprawl* and its associated development, and mining practices such as *strip mining.*

soil erosion   The movement by water or wind of *soil* (usually topsoil) from one place to another is called soil erosion. Erosion occurs naturally as runoff water flows into streams and rivers. The Mississippi River transports about 325 million metric tons of soil from the middle of America to the Gulf of Mexico every year.

Areas that get little rain on a regular basis or are prone to drought may experience wind-caused soil erosion instead of water-induced erosion. Most soils, however, are protected from erosion by plants that stabilize the soil. Serious erosion occurs when human activities remove most of the plant cover, exposing soil to elements. *Deforestation, clear-cutting,* poor farming practices, building construction, off-road vehicles, and other activities cause soil erosion. Even though soil formation occurs naturally, human activities, as mentioned above, remove it much faster than it is created. It takes about 200 to 1,000 years in most regions to create about 1 inch of topsoil, soil that can be eroded away in a few days. (See *Soil conservation, Dust bowl*)

soil horizons   Refers to the horizontal layers of soil. The upper horizons contain partially decomposed plant matter, called *leaf litter,* while the lower horizons have little or no organic matter. The bottom horizon, for example, consists of solid rock, called the *parent rock.* Each layer is created from the natural weathering of the parent rock and the organisms living in the region.

There are three main horizons, with each divided into more specific layers. The top layers form the A horizon, called the topsoil; the middle layers form the B horizon, called the subsoil; and the lower layers form the C horizon, composed of parent rock.

soil organisms   Soil provides nutrients and support for terrestrial plants to grow, but it also is a habitat teeming with life. A single gram of rich soil can contain: 2.5 billion *bacteria,* half a million *fungi,* 50,000 *algae,* and 30,000 protozoans (single-celled animals).

Topsoil acts as a transitional region between the living and the nonliving worlds. Minerals from the soil are absorbed by plants and become part of the living world. Many organisms break down the organic molecules in dead plants and animals, into simpler molecules so they can be re-used by other plants.

The soil community is hard to see but fascinating and complex. Bacteria, fungi, protozoans, nematodes, earthworms, and insects establish their own complex *food web.* Bacteria and fungi are major players in the decomposition of dead organisms. Some fungi attack and kill nematodes. Many protozoans appear to be *parasitic* on other soil organisms. *Nematodes,* also called roundworms, play many roles and are found in vast numbers in the soil. *Earthworms* condition the soil for plant growth. Although many soil *insects* are pests, many more are beneficial, acting as decomposers and soil conditioners.

soil particles   Soil particles are divided into three different categories based on their size. From largest to smallest, they are sand (2.0 to .05 millimeters), silt (.05 to .002 mm), and clay (less than .002mm). Particles larger than .05 mm are called *gravel.* Soils that contain a lot of sand feel gritty when wet. Soils with a lot of silt feel smooth like flour, and those with a lot of clay feel sticky when wet. (See *Soil texture*)

soil profile   Refers to a vertical section of soil that reveals the **soil horizons.**

soil texture   The soil texture is determined by differing mixtures of the three **soil particles:** sand, silt, and clay. For example, a soil with 40% sand particles, 40% silt particles, and 20% clay particles is called "loam." For a soil to be called "sand," it must contain at least 85% sand particles with the remaining portion a mixture of silt and clay. For soil to be called "clay," it must contain at least 40% clay particles and the remainder, a mixture of sand and silt.

soil types   There are thousands of soil types, but the most important soils can be grouped into two major categories: grassland soils called *chernozem* and forest soils called *podzol.* Grassland soil is formed in areas with limited amounts of rainfall. This keeps the nutrients in the top **soil horizon** ("A" horizon) from *leaching* down to the B horizon. Since the nutrients remain near the surface, root systems are shallow. Forest soils, however, occur in areas with more rainfall, resulting in nutrients leaching downward, producing a deeper B horizon, which supports large root systems.

solar cells See *Photovoltaic cells.*

solar ponds Solar ponds are unique power plants that use the sun's energy to heat water. They consist of black plastic bags filled with water, which are laid out to cover an area of at least one acre. The bags are held in a containerlike structure or a cavity carved into the earth. As heat is trapped within the water in the bags, it is used to produce steam, which in turn turns turbines to produce electricity. Israel is currently generating power with a solar pond near the Dead Sea and plans to build additional plants at the same site. Some analysts believe that solar ponds can supply significant amounts of pollution-free power in the future. (See *Hydroelectric power, Solar power, Moon power, Tide power, Wave power, Solar-thermal power plants*)

solar power At present, only 1 percent of the world's energy supply comes from solar power. Some scientists, however, believe the sun to be the ultimate **alternative energy** source and the answer to most of our energy problems. The sun delivers far more energy than all the peoples of the world could ever use, but the problem is harnessing this energy.

Four methods are currently used to harness the sun's energy: 1) **Passive solar heating systems** absorb the sun's energy as heat for immediate use. 2) **Active solar heating systems** use a solar collector device and a system of pipes to transfer the heat to a target area. Passive and active systems can heat homes and produce hot water. 3) **Solar-thermal power plants** use solar collectors to heat water or other liquid, which then generates electricity. 4) The **photovoltaic cell** converts sunlight directly into electricity.

Solar energy produces little or no air and water pollution, adds no carbon dioxide, a **greenhouse gas**, into the atmosphere, and doesn't destroy the land.

solar-thermal power plants Solar-thermal power plants use enormous mirrors to focus sunlight onto a central heat collection facility, which is composed of pipes containing either water or oil. The heat captured by the fluid in these pipes is used to generate electricity. Such a plant in the Mojave Desert (California) uses 1,000 acres of these mirrors and generates 80 **megawatts** of power, which is enough to meet the needs of a small city. The cost for this electricity is competitive with that of *fossil fuels* and cheaper than *nuclear power.*

These plants cause no air or water pollution and are relatively inexpensive and quick to build. (See *Solar power, Photovoltaic cells*)

source reduction Source reduction is a method of reducing the amount of *municipal solid waste* that usually ends up being dumped into *landfills* or *incineration* plants. Many people consider landfills and incinerators "Band-Aid fixes" for a much bigger problem—the production of too many disposable products.

Source reduction means minimizing the use of a material that will have to be disposed of at a later time. Source reduction of excessive packaging materials is one important area. Returnable plastic bottles and aluminum cans have been reduced by 20 to 35 percent of their weight since they were first introduced. CDs, which used excessive packaging so they could be seen on the older record racks, are voluntarily being reduced by the entertainment industry. Concentrated forms of detergent are being offered by many companies as alternatives to larger bottles containing diluted solutions. The volume of organic matter such as grass clippings, leaves, and leftover foods that would otherwise go into a landfill can be reduced by being separated and used to produce *compost.* (See *Waste minimization*)

space conditioning See *Thermal insulation, building.*

space debris pollution About 240 items per year are shot into space and placed into orbit to circle our planet. A recent report by the Congressional *Office of Technology Assessment* states that by the year 2000, satellites and other objects will become so cluttered as to pose a danger of a space debris collision.

speciation When *natural selection* (along with other factors) results in the development of a new *species,* the process is called speciation. This can occur when individuals of one species migrate to separate areas with different conditions. For example, some individuals belonging to an early species of fox are believed to have migrated north while others migrated south. Those that went north developed heavier fur (for warmth), shorter ears (to reduce heat loss), and a white coat to blend in with the snow, creating a new species called the arctic fox. Those that went south developed a thin coat, long ears, and a darker coat, establishing a new species called the gray fox.

The process of speciation can probably occur within a few hundred or thousand years with organisms such as insects that reproduce rapidly, but takes tens of thousands to millions of years in higher forms of life such as mammals.

species   A group of similar organisms capable of reproducing with one another. The species is the lowest (most specific) category of biological classification. (See *Kingdom, Speciation*)

speleology   The study of caves.

spoil banks   Refers to the resulting damage when *strip mining* is not restored.

stand   A group of trees with uniform characteristics such as all of the same species, age, or other condition. (See *Old-growth forests*)

standard of living   The standard of living refers to the quality of life. It is difficult to quantify since needs and values differ among different peoples. The United Nations has designated countries as *more developed countries (MDCs)* and *less developed countries (LDCs).* With few exceptions, these two designations can also be used to distinguish those countries with high standards of living versus those with low standards of living.

Certain indicators are often used to give a reading on a people's standard of living. The infant mortality rate is a good indicator of standard of living. In 1990, the MDCs averaged 16 deaths per 1,000 births while the LDCs averaged 81 deaths per 1,000 births. (The U.S. rate was 10.) Life expectancy is another good indicator. In 1990, the MDCs averaged a life expectancy at birth of 74 years while the LDCs could expect 61 years. (The U.S. rate was 75 years.) (See *Demographic transition*)

standing crop   Refers to the amount of living matter in an ecosystem at any given time. (See *Biomass, Net primary production*)

standing water habitats   Freshwater ecosystems can be divided into those found in *running water habitats* and those found in standing bodies of water such as ponds, lakes, and reservoirs. If a standing body of water is deep enough, it has a top layer, called the euphotic zone, in which sunlight penetrates, and a lower level that receives no light, called the aphotic zone.

In areas where the light reaches the bottom of the water, rooted plants grow. This part of the euphotic zone is called the littoral region. It may contain flowering plants such as water lilies and cattails. These are called emergent plants since they emerge from the water. Aquatic plants that remain entirely submerged are called submergent plants and include elodea.

Algae also play an important role in these habitats. They form thick layers on the bottom or at the surface attached to vegetation. The emergent and submergent plants, plus the algae masses, provide food and shelter for many species of fish, insects, crayfish, clams, and other aquatic organisms.

If the body of water is deep, the lower portion of the euphotic zone has no rooted plants and is called the limnetic region. These regions are in many ways similar to *marine ecosystems. Phytoplankton* float in the sunlit area, performing photosynthesis and becoming the first link in many food chains. *Zooplankton* feed on the phytoplankton, which are eaten by small fish, which in turn are eaten by *predatory* fish.

Since the lower levels (aphotic zone) receive no light, inhabitants must get their nutrients from dead and decaying organisms and waste that floats down from the surface regions, or they may travel to the surface to find food. The organic matter that continually falls to the bottom produces a nutrient-rich environment. Organisms that dwell in the organic ooze that accumulates at the bottom are mostly anaerobic bacteria, since there is little oxygen at this level.

Many terrestrial and flying organisms are associated with these standing water ecosystems, including dragonflies, turtles, frogs, mallards, muskrats, and others. (See *Eutrophication, Water pollution*)

Statement of Forest Principles   This document is one of five written for debate at the *Earth Summit* in June 1992. There are 15 principal elements in this document. Some elements of this document were controversial. The following portion was hotly debated: "States have the right to utilize, manage, and develop their forests in accordance to their development needs and level of socioeconomic development." The statement also supports the need for financial resources to developing countries with significant forests that are establishing programs for their conservation. This document is not a legally binding statement.

steppe See *Grassland.*

strip cropping **Soil erosion** can be reduced by alternating rows of one crop, such as corn, with a cover crop, such as alfalfa. This is a **soil conservation** technique called strip cropping. The alfalfa covers the soil and protects it from erosion by catching runoff from the alternate crop (corn, in this case). This method also helps minimize the spread of pests and plant diseases since there is no continuous stand of a single crop. The cover crop can be a nitrogen-fixing plant, which also helps restore the soil's fertility. (See *Crop rotation, Nitrogen fixation*)

strip mining Coal is mined from the earth by one of two methods: strip mining or **subsurface mining.** There are two kinds of strip mining (which is also called surface mining). The first, called area strip mining, is used on flat terrain where the vein of coal is no more than 300 feet beneath the surface. The material above the coal is called the overburden. Heavy equipment removes the overburden and places it alongside the vein. The vein of coal is then removed and the overburden replaced to fill-in the vein.

The second method, contour strip mining, is used on hilly terrain. In this method, the hill is cut into a series of terraces with the overburden from one dropping into the excavated terrace below.

In both methods of strip mining, the land must be manually restored (reclaimed) or **soil erosion** destroys the region. In area strip mining, failure to restore the site results in a series of small eroded hills and valleys called spoil banks. In contour strip mining, a wall forms that is quickly eroded, called a highwall. (See *Acid mine drainage*)

styropeanuts The popular packaging "peanuts" you often find pouring out of a shipping box are called styropeanuts. Just like the foam coffee cup, they are made from **polystyrene (PS),** which accounts for about 11 percent of all plastics. Since they are inherently bulky, they take up a great deal of space in **landfills** and, like most **plastic** products, are nonbiodegradable. In an effort to resolve the disposal problems caused by styropeanuts, a number of options are evolving.

Recycling projects have started recently. Some small-scale projects reuse styropeanuts in their original form. A group of chemical companies plan to open five polystyrene recycling plants, which will clean and pelletize these products so they can be used to manufacture items such as trays and wastebaskets. They hope to be recycling 25 percent of the polystyrene by 1995.

Alternatives to styropeanuts are also springing up, including the use of popcorn, which is, of course, biodegradable. A new generation of these packing "peanuts" is now produced out of vegetable starch. This foamlike "peanut" is nontoxic and degrades on contact with water. (See *Plastic recycling*)

subsurface mining   When coal is too deep for **strip mining,** subsurface mining is used. A deep vertical tunnel is dug, and a series of tunnels and rooms are blasted out so the coal can be hauled to the surface. The room-and-pillar method leaves about half the coal in place to act as pillars, which provide structural support to the maze.

succession   While organisms live in their environment, they change that environment, making it less suitable for themselves and more suitable for other types of organisms. This process results in an area being inhabited by collections of different organisms in predictable stages, with the simplest forms of life present early and more complex forms, later. This series of changes in the community is called succession.

The first stage is called the pioneer community, and the final stage that remains stable with little further change is called the climax community. All the stages together are called a sere. Different kinds of climax communities are called **biomes.**

There are two major kinds of succession: primary and secondary. Primary succession occurs in an area that has never been colonized by organisms, such as with barren rock created from a lava flow. This is called terrestrial primary succession. When the environment is a water habitat, it is called aquatic primary succession.

Secondary succession, which is more common, occurs when an existing community has been totally destroyed by events such as forest fires, floods, or clear-cutting a forest. Secondary succession differs from primary succession since the land has not been reduced to bare rock. (See *Succession, primary terrestrial; Succession, primary aquatic*)

succession, primary aquatic   All ponds and lakes are destined to become part of the land. This natural progression is called primary aquatic succession. It assumes the process has been going on ever since the body of water was created and assumes it will continue until the body of water is completely filled in. The entire process may take thousands of years.

The basic processes are similar for all types of *succession,* but with aquatic environments there is a continual flow of soil and organic matter from the surrounding land into the water. As the nutrient-rich materials enter the water, more organisms establish themselves. When they die, they contribute even further to the increase in sediment and organic matter at the bottom. As the water becomes more shallow, new species of plants that emerge out of the water establish themselves.

Once the sediment makes the body shallow so plants thrive near the surface, the water begins to dry out, creating a wet meadow. Once the body is dry, terrestrial organisms move in, with the process continuing until a climax community is established. (See *Succession, primary terrestrial; Aquatic ecosystems*)

succession, primary terrestrial A barren terrestrial region will become inhabited by simple organisms at first and then gradually give way to more advanced organisms until a stable climax community is developed. The process of gradual change in the makeup of a community is called *succession* and usually takes several hundred years to complete. When the process begins with barren rock, it is called primary terrestrial succession. (If it began with standing water, it is called primary aquatic succession.)

In the past, large portions of the earth were left barren when the glaciers receded at the conclusion of the last *ice age.* Today, barren areas are occasionally produced naturally due to volcanic action creating new lava formations, large-scale mud slides, the creation of new sandbars, or other catastrophic events. Barren regions are also created by human processes such as strip mining, which denudes the earth's surface.

The successional stages of primary succession, beginning with bare rock, are as follows: The *pioneer community* includes lichens that help establish a shallow layer of soil by breaking down the rock (along with erosion) and forming *humus* when they die and decompose. The *soil* allows mosses, some *annual plants,* small worms, and *insects* to establish themselves. As the soil becomes richer due to the decomposition of the organisms, the intermediate stages of succession occur. This includes the establishment of grasses, shrubs, and often some shade-intolerant trees, along with larger animals.

The final climax community begins to establish itself; it may include shade-tolerant trees and a full range of animals. The climax community is determined by many factors, including the temperature range and amount of water for the area. Succession may begin in some areas with sand instead of rock. (See *Biomes*)

succession, secondary   Secondary succession is similar to primary succession except for the initial condition of the environment. In primary succession, the land is barren and has never been colonized. Secondary succession begins when an area is devastated by some natural or manmade events. Examples of natural occurrences include forest fires and floods, while artificially produced events include **clear-cutting,** intense **water pollution, mineral exploitation,** or clearing woodlands for farms (and then abandoning the farm).

Since soil or sediment and organic nutrients still exist in these environments, the speed with which succession occurs is more rapid than that of primary succession, taking one or two hundred years.

An example of secondary succession on abandoned farmland in the southeastern U.S. goes through the following stages, beginning with a plowed field. The pioneer stages involve the establishment of annual weeds such as crabgrass. This occurs during the first couple of years, when the dormant seeds that lie in the soil germinate. Over the next decade, grasses and other small **perennial plants** take hold along with a few shrubs. Sedge is a **dominant** species during the early stages. The sedge stabilizes the soil enough for pine and spruce seedlings to take root. Over the next 10 to 20 years the pine trees have grown tall enough to shade the ground, eliminating the sedge. The succession has now moved into the intermediate stages.

Competition for moisture and sunlight prevents new pine seedlings from surviving among the existing **stand** of trees. Hardwood trees such as oak begin to grow and force out the pines. Finally, shade-tolerant trees such as dogwood fill the **understory** in the climax stage, resulting in a relatively stable community. (See *Biomes, Succession*)

sudd   A large, floating mass of plants that can clog streams, rivers, and dams.

sulfur dioxide   Sulfur dioxide is one of five primary pollutants that can cause **air pollution.** It is released when **fossil fuels** containing sulfur are burned. Sulfur is often present in fossil fuels because it was naturally found in organisms during **coal forma-**

tion and *oil formation.* Burning oil or coal that contains sulfur produces an odor and causes respiratory irritation. It can react with oxygen and moisture to produce acids (*secondary air pollutants*) that can destroy lung tissues and contribute to *acid rain.*

Industrial smog (also called gray-air smog) contains sulfur dioxide. Modern coal- and oil-burning facilities often use "scrubbers" and other sulfur control devices that dramatically reduce these emissions. Some plants in the U.S. and especially less developed countries still generate sulfur dioxide in quantities harmful to humans.

The best way to reduce sulfur emissions is to convert from burning high-sulfur content coal to low-sulfur content coal, oil, or natural gas, which reduces emissions by at least 65 percent. Another method is to remove the sulfur from the coal before it is used, but this process is expensive and not often used. (See *Coal formation, Oil formation*)

sunflower oil, diesel fuel   Sunflower, soy, and canola oil are being tested as an alternative to automobile diesel fuel. Supposedly, no modifications are necessary to the engine. This type of fuel is being produced in Italy under the name of Diesel-Bi, which is 90 percent oil and 10 percent methanol. Environmentally, there are many advantages over conventional fossil fuels. Since these oils don't contain sulfur, they don't produce *sulfur dioxide,* which contributes to *acid rain.* Independent tests have indicated that this fuel reduces soot emissions by 80 percent, carcinogens by 65 percent, visible smoke by 65 percent, and *carbon monoxide* by 30 percent. (See *Alternative energy*)

sunsetting   Refers to the gradual phase out of production and use of toxic substances. For example, sunsetting mercury could begin with its banishment from batteries and paints, followed by its removal from many other products and processes until it is totally eliminated. (See *Toxic waste, Heavy-metal pollution*)

super-insulation   Refers to insulation used in buildings constructed with a variety of the most advanced insulating technologies including features such as double- or triple-glazed, gas-filled, low emissivity windows and heavy insulating fiberglass batts with *R-values* of 35 to 40. Super-insulation dramatically reduces energy costs. (See *Thermal insulation, buildings; Chalking, insulation*)

Superfund In an effort to clean up the numerous **hazardous waste disposal sites** located throughout the U.S., Congress established CERLCA (Comprehensive Environmental Response, Liability, and Compensation Act), which set aside a large trust fund, commonly called the Superfund, to be managed by the **EPA.** This fund is used to clean up those hazardous waste sites considered most dangerous and placed on the National Priority List. The act was also designed to establish a comprehensive cleanup plan and make those responsible for the pollution pay for most of the cleanup. This act has been updated and funds increased, but it is still considered by many environmentalists to be more smoke than substance. Of the 1,200 + sites on the list, fewer than 60 have been cleaned-up to date.

One of the reasons for this poor track record is that offenders find it more economical to fight the EPA in court than to clean up the toxic wastes. Almost five billion dollars of the Superfund budget has been spent on legal costs instead of cleanup costs.

surface mining See *Strip mining.*

surface water The world's freshwater resources are divided into surface water and **groundwater.** Surface waters include streams, rivers, ponds, lakes, and manmade reservoirs. All fresh water that is not absorbed into the earth (becoming groundwater) or returned to the atmosphere as part of the **water cycle** is considered surface water. Only about .02 percent of all **water** on our planet is surface (fresh) water.

sustainable agriculture Many agricultural practices used today contaminate the soil and water with synthetic **fertilizers** and **pesticides. Soil erosion, salination,** and **waterlogging** are all harmful side effects of today's agriculture. Short-term productivity is paid for by long-term destruction of the land. Sustainable agriculture tries to establish an ongoing relationship with the land, resulting in long-term productivity with few harmful effects.

Sustainable agriculture uses the latest advances in technology and old-fashioned farming practices to assure continued use of the land and less harm to the environment. Ongoing research in the field of sustainable agriculture is not a "cute" idea, but one that is economically feasible for the farmer as well as environmentally sound for everyone.

Many land grant universities have established "low input sustainable agriculture" programs with federal research money. Many components of sustainable agriculture are similar to **organic farming** techniques, but on a larger scale. They include a combination of old farming practices such as **crop rotation** and **strip farming** and new technologies such as genetically engineered crops that resist pests and drought. **Integrated pest management** is also used in sustainable agriculture to reduce the use of pesticides.

sustainable biosphere initiative (SBI) The SBI is a major effort put forth by the Ecological Society of America to define research priorities for **ecology** throughout the 1990s. This initiative focuses on the role the ecological sciences will play in the wise management of the earth's resources and maintenance of life-support systems. The Executive Summary of this document calls SBI a "call-to-arms" for all ecologists. Write to The Ecological Society of America, Public Affairs Office, 2010 Massachusetts Avenue NW, #420, Washington DC 20036.

symbiotic relationships When two organisms of different species live together over long periods of time in close physical contact, with one or both benefiting in some way, it is called a symbiotic relationship. Symbiotic relationships can be divided into four categories: parasitism, **commensalism, mutualism**, and **parasitoidism.** (See *Parasite, Predator-prey relationship, Food webs*)

synecology Synecology (also called community **ecology**) studies how groups of **populations,** called a **community**, coexist in the same environment at the same time. It studies why the number and mix of different species in a given area change over time, what controls the dispersion of a population in an area, and why these changes occur. (See *Migration, Competition*)

synfuels Solid **coal** that has been converted into either a gas or a liquid fuel is called synfuel, for synthetic fuel. **Coal gasification** converts solid coal into a gas synfuel called synthetic natural gas (SNG or **syngas**). Coal liquefaction turns it into a liquid synfuel such as methanol. Both types of synthetic fuels produce much less air pollution than burning solid coal. Liquid fuels are more functional than solid fuels for heating homes and running automobiles and other forms of transportation. They can also be transported through pipelines whereas solid coal must be shipped.
    Synfuel facilities, however, are expensive to build and run com-

pared with a coal-burning power plant that has a full complement of air pollution control devices. Synfuels also have a lower heat content, meaning more coal must be used to produce the same amount of energy from synfuels than directly from coal.

syngas   Refers to a gaseous synthetic fuel (**synfuel**) created from either **coal** or **biomass** by the process of gasification. The name is short for synthetic natural gas since the substance is similar to natural gas but has anywhere from one-quarter to the same amount of available energy as natural gas. (See *Biofuels*)

synthetic natural gas (SNG)   See *Synfuels*.

systems ecology   This branch of ecology studies the relationships between all the components of an ecosystem, living and non-living, and concentrates on studying how energy and chemicals flow through the system. The term "system" can refer to simple biological groups such as **habitats** (a small pond or a forest), or as large as the entire **biosphere. Computer modeling** is often used to predict how an ecosystem might behave to changes in certain factors such as increased concentrations of pollutants or gases in the air.

tagging programs   Tagging or marking animals is a useful and accepted method of monitoring and studying wildlife. Everything from tagging fish, banding bird legs, attaching radio transmitters on grizzly bears, and using delicate mylar patches on monarch butterflies have been used successfully. These studies help scientists study growth rates, migratory paths, and breeding cycles. The geographical range of the polar bear, wintering grounds of the monarch butterfly, and the breeding grounds of some marine turtles have all been discovered through the use of tagging programs.

taiga   Taiga is one of several ***biomes.*** The primary factors that differentiate biomes are temperature and precipitation. The taiga is also called a northern coniferous forest or a boreal forest. Precipitation is between 25 and 100 centimeters (10 to 40 inches) per year. These regions have long, harsh winters with short, cool summers.

The ***dominant*** plants are coniferous trees, including spruce, fir, and larch with needlelike leaves that minimize water loss. (Water is precious since it is locked up in snow and ice through the long winters.) Most birds are migratory, and most of the insects become inactive in the winter. Many small mammals inhabit the taiga, and larger animals include deer, moose, and wolves.

tailings, uranium mining   After uranium ore is mined for use in ***nuclear reactors,*** it must be crushed for processing. The remaining piles of crushed rock are called tailings. They release low-level ***radiation*** and pose health risks. Radioactive dust can be blown by the wind away from the original tailings pile and contaminate areas far from the source. Radioactive particles can also leach through the soil and enter the groundwater or simply be washed into rivers or streams, contaminating them.

Mine tailings (from radioactive and nonradioactive minerals) usually reject plant growth resulting in soil erosion. (See *Mineral exploitation*)

target organism   ***Pesticides*** are designed to kill a specific pest, which is called the target organism. Any organism affected, other than the target organism, is called a nontarget organism. (See *Pesticide dangers*)

tarn A small lake or pond located high in a mountain region.

taxol Taxol is a natural substance found in the bark of yew trees that grow scattered throughout the ***ancient forests*** of the Pacific Northwest. Studies have shown taxol to be an effective treatment for some advanced forms of cancer. Very little taxol can be extracted from these trees. Six 100-year-old trees produce enough taxol to treat only one patient. Attempts are being made to extract the drug out of the tree needles, which can regrow, as opposed to stripping each tree of bark, killing the tree, removing it as a source of further taxol. In addition, scientists are trying to synthesize the drug, so it can be manufactured in the laboratory. (See *Deforestation*)

taxonomy The study of describing, naming, and classifying organisms.

telephone pole disposal There are about 120 million telephone utility poles in the U.S. with over 2 million needing replacement each year. To prevent decay and fire, the poles are treated with chemicals such as creosote, which the EPA considers a toxic substance and is potentially ***carcinogenic.*** Recent studies and test programs have revealed bacteria that thrive on the wood and the impregnated toxic chemicals. These ***bacteria*** are being tested to render the wood harmless so it can be recycled to make wood products such as desks and other types of furniture. (See *Recycling, Toxic wastes*)

temperate deciduous forest A type of ***biome.*** The factors that differentiate biomes are temperature and precipitation. These forests receive over 100 centimeters (40 inches) of rain evenly distributed throughout the year. There are cold winters in which the plants lose their leaves and become inactive until the following spring. Although life is abundant, most of these forests are composed of only a few species of trees. The forest floor contains many flowering plants that have access to the sunlight in early spring before the leaves return to the trees. There are also many small shade-tolerant shrubs.

These regions contain numerous insects, small mammals such as mice and rabbits, and large grazing mammals such as deer. ***Predators*** such as foxes and badgers are also common. Most of the birds that live in these woods migrate south during the colder months.

Since the soil in these regions is so fertile (primarily from the loss of leaves each season), many of these forests have been cleared and converted to farmland; others have been swallowed up by **urban sprawl.** In North America only one-tenth of one percent of the original forests remain.

teratogens   Refers to substances that cause birth defects. (See *Toxic waste*)

terracing   Crops grown on steep slopes are prone to excessive **soil erosion.** Terracing is a **soil conservation** technique in which the slope is carved into a series of level terraces where the crop is grown on each terrace. Water gradually flows from one terrace to the next in a cascade.

terrestrial ecology   The study of organisms that live on land and their habitats. (See *Ecology*)

theoretical ecology   Theoretical ecology is the latest approach to the study of ecology. It uses mathematical equations and **computer modeling** on powerful computers to predict what will happen to populations of organisms and life on earth. These equations and models are based on existing knowledge, but must also rely on many assumptions. A typical example is the attempt to predict what will happen to life decades from now if **global warming** or **deforestation** continues at current rates.

thermal insulation, building   Fifty to 70 percent of the energy used in the average American home is used for space conditioning (heating and cooling). Proper thermal insulation dramatically reduces the amount of energy used for space conditioning by minimizing air leakage. The three most important areas requiring insulation are: the attic space, under floors that are above unheated spaces, and around walls in basements and crawl spaces.

Thermal insulation is measured by the **R-value,** which indicates the resistance to heat flow. The Department of Energy has recommended R-values for each zip code in the country. For new-construction buildings, be sure the recommended R-values are met throughout the home. For existing construction, insulation can almost always be upgraded to meet these recommendations.

There are many different types of insulation currently used. These include blankets or batts of mineral fibers, loose or blown fill made of fiberglass, rock wool or cellulosic fiber, or perlite or vermiculite. Each type of insulation requires a different thickness to reach the desired R-value. For example, 9 to 10 inches of blown-in rock wool offers an R-value of 30, but you'll need 13 to 14 inches of fiberglass to attain the same R-value. (See *Super-insulation*)

thermal inversion   Pollutants produced within cities are usually dissipated as warm air at the surface rises and mixes with cooler layers above. Cities within valleys or partially surrounded by mountains are prone to thermal inversions in which a layer of warm air forms above the surface layer, forming a lid over cool air. This traps pollutants for extended periods of time, often resulting in **photochemical smog.** Thermal inversions and photochemical smog are often problems in Los Angeles, Salt Lake City, Phoenix, Denver, and Mexico City. (See *Dust dome, Urban heat island*)

thermal water pollution   One form of **industrial water pollution** is called thermal water pollution. Two hundred and twenty-five billion liters of water are used each year by industry, with most of it used in power plants that generate electricity. These power plants convert water to steam, which drives turbines to produce electricity. The steam is then cooled by more water, which carries the heat away. When the water is released into natural bodies of water such as a river or stream, **aquatic ecosystems** are affected, since all organisms have specific temperature and dissolved oxygen **tolerance ranges** in which they can survive. (Temperature controls, in part, the amount of dissolved oxygen in the water.) Changing the temperature just a few degrees, which often happens, can change which organisms can and cannot survive.

Methods exist to reduce the water temperature before being released into natural bodies of water, such as using manmade artificial ponds to dissipate the heat. (See *Water pollution*)

thermochemical conversion, biofuel Biomass **energy** is produced by either **biochemical conversion** or thermochemical conversion. The latter produces methanol fuel or synthetic natural gas (**syngas**). The process involves heating some form of biomass, such as wood, in air containing little or no oxygen. This results in the biomass breaking down into simpler substances, which can be used as fuel. **Gasification** is one method of thermochemical conversion that converts biomass (usually waste wood products such as sawdust) into syngas, which can be used for fuel.

Thermodynamics, Second Law of   This law states that when energy is converted from one form to another, some of the useful energy is lost, usually in the form of heat. When coal (chemical energy) is burned to create electricity (another form of energy) in our power plants, large amounts of energy are lost as heat and become useless. The same loss of energy occurs between each trophic level as illustrated in an *energy pyramid.*

THM This is an acronym for trihalomethanes. These are substances produced when water is chlorinated as part of a municipality's drinking-water treatment process. The chlorine gas comes in contact with organic matter floating in the water such as leaves, grasses, or food particles, and produces THMs. Chloroform is a well-known THM. Many THMs have caused cancer in laboratory animals and may contribute to birth defects. For this reason, the EPA has set an allowable limit of 100 parts of THMs per billion parts of drinking water. Since many treatment plants have trouble meeting this standard, many are beginning to use alternatives that don't produce THMs. (See *Chlorination*)

threatened species Threatened species (often called vulnerable species) refers to species that are likely to become *extinct* if some critical factor in their environment is changed. In other words, the species now exists with some environmental factor at the minimum level sufficient for survival. One such animal is the *Northern Spotted Owl,* which has lost most of its habitat to logging. (See *Endangered Species Act*)

Three Mile Island accident (TMI) The worst nuclear power plant accident in the U.S. occurred on March 29, 1979, at the Three Mile Island nuclear power plant in Pennsylvania. A series of equipment failures and human errors resulted in the reactor losing its coolant and having a partial *meltdown,* resulting in an unknown amount of radiation loss. An evacuation advisory was issued. It cost over one billion dollars to close the reactor. (See *Chernobyl, Nuclear power*)

three R's The environmental version of the three R's stands for Reduce, Re-use, and Recycle. (See *Environmental education, Environmental literacy, Source reduction, Recycling*)

throwaway age A phrase coined by Vance Packard in his 1960 book, *The Waste Maker,* describing a society in which everything is disposable. (See *Landfills, Recycling, Plastic pollution*)

tires, recycling   Over 240 million tires are discarded in the U.S. each year. Currently, about 7 percent are recycled, 11 percent are incinerated and used as a fuel source, and about 5 percent are exported. That leaves almost 80 percent that must be disposed of in landfills or stockpiled. It is estimated that about 2 billion old tires lie in rubber mountains across the country. New technologies are being developed that can create viable markets for most of these tires and relieve the burden placed on landfills. Some uses include reprocessing and recycling the rubber for paving materials and rubber flooring materials.

Title X of the Public Health Service Act   This is the only U.S. federally funded program devoted to family planning. Its budget was dramatically cut between 1980 and 1992. (See *Doubling-time in human populations*)

tolerance range   Pertains to the range of a specific factor, such as temperature, in which an organism can survive. If the factor goes beyond the tolerance range (too high or too low), the organism will die. This then becomes the **limiting factor.** (See *Habitat*)

Tongass National Forest   The Tongass National Forest is the largest of all the 156 national forests in the U.S., containing almost 17 million acres. It makes up about 80 percent of southeast Alaska and is 57 percent temperate rain forest. It contains trees 800 years old and 200 feet tall and has the world's largest concentration of bald eagles and grizzly bears.

The U.S. **Forest Service** signed 50-year contracts with two large logging firms to log over 90 percent of this forest. During the late 1980s, environmentalists brought to the public's attention the loss of nature and the loss of money these contracts had perpetrated on the public. These groups tried to get Congress to reform the U.S. Forest Service's management of Tongass as well as other national forests. In 1989, Congress voted to halt timber sales within Tongass. (See *National Park and Wilderness Preservation System*)

topsoil   Refers to those layers of soil that contain large amounts of **organic** matter. (See *Soil horizon, Soil organisms*)

touchpools   Public aquariums, like zoos, attract millions of visitors each year. Touchpools in an aquarium are the equivalent of the petting farm section of a **zoo.** Aquatic or marine organisms that can be safely handled by visitors are placed into a shallow pool accessible to the visitor. Guides are usually present to urge visitors to handle the specimens and explain about their natural history.

toxic pollution   Toxic pollution refers to any substance introduced into the environment that causes harm to an organism's normal functioning. Since these substances can be found in the air, water, or soil, toxic pollution is usually a component of **air pollution** and **water pollution.** In the U.S. alone, millions of tons of toxic pollutants are produced and disposed of as **toxic wastes** each year. The government has passed legislation in response to the public's concern about toxic pollution, including the **Resource Conservation and Recovery Act,** the **Toxic Substances Control Act,** and the **Superfund.**
  Technological advances are enabling industry to use less of these substances without having a negative effect on their business and competitiveness. Cleaning up existing toxic wastes and **hazardous waste disposal** sites may prove more difficult than preventing future damage.

Toxic Substances Control Act of 1976   This piece of legislation gave the **EPA** the authority to require testing of new chemicals before they can be made available to the public. (See *Pesticide dangers, Toxic pollution*)

toxic waste   Toxic waste is one type of **hazardous waste.** Toxic waste consists of substances that negatively affect the health of organisms. Some toxic substances are harmful or lethal if an individual is exposed to a single large dose. This is called **acute toxicity.** The chemical released at the plant in **Bhopal** was acutely toxic. Other toxic substances are harmful in low doses over long periods of time, such as mercury and lead. This is referred to as **chronic toxicity.** These substances can cause harm because they are carcinogens (cause cancer), teratogens (cause birth defects), or mutagens (cause immediate genetic defects). (See *Toxic pollution*)

toxicology   Pertains to the study of poisons and their effects on organisms.

trace elements   Elements that occur in organisms in minute amounts, but are essential to life; examples are copper and zinc. (See *Nutrients, essential*)

tract development   A type of **urban sprawl** in which homes or other residential units are built over large areas, usually on land originally used for farming. (See *Ribbon sprawl, Farmland lost, High-market value farming counties*)

tragedy of the commons   Refers to the overuse of a common resource resulting in the degradation or depletion of that resource. For example, fish are often harvested until the fishing grounds are empty, or land is irrigated until the groundwater is gone. The tragedy is defined by the belief that if "I don't use it, someone else will." The phrase originated in biologist Garrett Hardin's essay "The Tragedy of the Commons." In the essay, the commons was pasture land; today, however, it can be extended to include the entire biosphere. (See *Natural resources, Mineral exploitation*)

transpiration   Refers to water loss from plants. (See *Water cycle*)

Treaty on Biological Diversity   This treaty signed at the **Earth Summit,** held in Brazil in 1992, requires participating countries to take inventories of plants and animal species and continue to protect endangered species. It also requires participating countries to share the research, profits, and resulting technologies with countries whose resources are used in the research. The United States was the only country at the summit that didn't sign this treaty.

tree   A large woody plant, usually with a single trunk that is at least 3 meters (10 feet) tall. (See *Bush*)

trophic levels   See *Energy pyramid.*

Tropical Forest Consumer Information and Protection Act   This act was proposed by members of Congress to help reduce tropical **rain forest destruction.** The act requires labels on wood products containing woods imported from **tropical rain forests.** (See *Deforestation*)

tropical grassland   See *Savanna.*

tropical rain forest   Tropical rain forests are one of several kinds of **biomes.** Tropical rain forests receive 200 to 500 centimeters (80 to 200 inches) or more of rain per year. They are found near the equator, where the temperatures are always warm. The *biodiversity* of these biomes is truly amazing. There may be hundreds of species of trees living in close proximity. There can be three or four layers of *canopy* due to this diversity. The canopies are thick and provide a unique habitat that can be studied as an entire *ecosystem* unto itself. The plants living on the ground must be shade-tolerant since little light filters through all the trees. Ferns, moss, flowering plants, and vines are also abundant.

Animals include numerous birds, rodents, insects, snakes, and lizards. Many animals are tree-dwellers ranging from frogs to mammals. Even though tropical rain forests take up a fraction of the earth's surface, they contain more plant and animal species than all the rest of the planet combined. About 25 percent of all pharmaceuticals are derived from plants that grow in tropical rain forests. These regions are being harvested for timber, mined for minerals, and cleared to make way for crops and rangeland for grazing animals at staggering rates. (See *Deforestation, Rain forest destruction*)

tropics   The region of the earth lying between the Tropic of Cancer and the Tropic of Capricorn, which are the northern and southern limits, respectively, to where the sun appears directly overhead at noon. Pertains to warm climates.

tub bath vs. shower   A typical tub bath uses about 30 gallons of water. A typical shower runs at about 6 gallons per minute. This means a 5-minute shower uses about the same amount of water as a tub bath. A 10-minute shower uses 30 more gallons of water than a tub bath (60 gallons). Shower head flow controllers or low-volume shower heads can reduce the volume of water used in the shower substantially to save water. (See *Domestic water use, Domestic water conservation*)

tundra   Tundra is one kind of *biome.* Tundra (also called polar grasslands) gets about the same amount of precipitation as deserts (25 cm or less, about 10 inches), but the conditions are very different. Tundra is characterized by a permanently frozen layer of soil called permafrost. These regions occur at the extreme northern latitudes, north of the *taiga,* and at high altitudes where similar conditions exist, called alpine tundra.

During the summer, there are a few months in which just the top of the soil thaws and a few small types of plants quickly grow. Since the ground is always frozen, water remains at the surface, creating many shallow bodies of water. Migratory birds fly in for a brief period. Insects become prevalent in the brief summer also. A few hardy mammals can survive this environment, including mush oxen, caribou, and the arctic hare.

turtle excluder device (TED)   Shrimp trawlers in the Gulf of Mexico and off the southeastern shore of the U.S. trap roughly 45,000 unwanted sea turtles each year in their long, cone-shaped shrimp nets. Over 10,000 of these turtles die of suffocation since they are trapped beneath the water for long periods of time. Recent federal legislation mandates the use of turtle excluder devices (TEDs), which allow almost all of these creatures to escape from the net (without releasing the shrimp).

The TED is a metal or nylon grid that is inserted into the middle portion of the trawl cone. The grid is large enough to allow shrimp to pass through and remain netted. Large organisms, however, such as sea turtles, bump into the grid and are stopped. The grid is placed at an angle, so the turtle slides toward the side of the net and is released through an opening.

Shrimpers argue that these devices make them less competitive than Mexican shrimpers (who are not required to use them) and will harm the U.S. shrimp industry. (See *Driftnets, Gillnets, Purse seine nets, Marine Mammal Protection Act*)

ubac Refers to the northern, shady side of a mountain, which contains fewer trees and has a lower snow line.

understory Refers to the layer of vegetation between the **_canopy_** and the groundstory in a forest; usually formed by shade-tolerant trees of medium height. (See _Temperate deciduous forest, Tropical rain forest_)

Union of Concerned Scientists This organization was founded in 1969 for scientists to participate in important decisions regarding technology's impact on society. There are two programs: energy and arms. The energy program deals with **_global warming_**, national energy policy, transportation, and **_nuclear power_** safety. The arms program focuses on nuclear weapons, arms control, and nuclear proliferation. Both programs produce books, reports, brochures, videos, and an award-winning quarterly publication called _Nucleus_. Write to 26 Church Street, Cambridge, MA 02238.

United Nations Environment Programme (UNEP) The UNEP was conceived at the 1972 Stockholm Conference, and is considered the environmental conscience of the United Nations. Its primary function is to raise the level of environmental action and awareness on a worldwide basis. This program is composed of 1,000 programs and projects that cover all aspects of the ecology of the planet. It coordinates all the United Nation agency's environmental activities and monitors global environmental quality. It publishes numerous papers, reports, books, and educational materials for worldwide distribution. Write to UNEP, DC2-0803, United Nations, New York, NY 10017. (See _Earth Summit, Global Environment Monitoring System_)

United Nations Population Fund This U.N. fund supplies about 140 **_less developed nations_** with family planning, maternal and child health care. The U.S. contribution to this fund was dramatically cut between 1982 and 1992.

upwelling   Off the western coasts of many continents, constant trade winds blow offshore. This pushes the water at the surface away, which in turn is replaced with deeper water that rushes up to fill the void. This process is called upwelling. The floor of most coastal regions of the ocean (**neritic zone**) receives sunlight and therefore is teeming with life and contains nutrients. When the cold deeper water rushes upward, it brings with it these nutrients. The nutrients help establish complex food chains by feeding plankton, fish, and predatory seabirds.

Upwelling occurs in only .1 percent of the total marine environment, but contributes significantly to the overall productivity of the ocean. These areas are also important for humans since many commercial fisheries are found in these areas. These include the tuna catch off the West Coast of the U.S., the anchovy catch off of Peru, and the sardine catch off of Portugal. (See *Oceans*)

Urban Carbon Dioxide Project   In 1991, an international consortium of cities, including Denver and Minneapolis in the U.S., was established to help cut **greenhouse gas** emissions by 20 percent in each of the member cities.

urban heat island effect   Cars, factories, furnaces, and people in urban areas generate large amounts of heat. All the asphalt, concrete, steel, and other construction materials absorb and retain vast amounts of heat. This tendency to generate and absorb heat causes cities to be 5 to 10 degrees F warmer in the summer than the surrounding countryside. This is called an urban heat island effect. This heat often creates a bubble over the city with its own wind currents and microclimate that trap pollutants, increasing the level of air pollution. (See *Dust dome*)

urban open space   Areas in urban settings that have been set aside for outdoor recreation are called urban open spaces. Since this property is often much more valuable if it were to become developed, city planners must see the intrinsic need for these spaces and their value to the city residents. Some of the world's biggest cities planned for open spaces while still young. For example, New York's Central Park was created in the late 1800s and consists of 500 acres of some of the world's most valuable real estate. (See *Nature centers*)

urban sprawl   Development of areas, surrounding large cities, that are less densely populated and often more environmentally pleasing, is called urban sprawl. As urban sprawl spreads between large cities, a large, highly populated area called a **megalopolis** is often formed. (See *Ribbon sprawl, Urbanization*)

urbanization and urban growth Urbanization refers to an increase in the percentage of individuals that live in urban areas. (Urban areas are villages, towns, or cities with populations greater than 2,500.) Urban growth simply refers to an increase in the urban population size. Urban areas have increased, on a worldwide basis, from only 14 percent of the total population in 1900 to 43 percent in 1985. People move to urban centers for many reasons, including the shift from a rural farming society to an industrial society, meaning the jobs are in the city.

Urbanization has come with a reduction in the standard of living for many individuals throughout the world, but especially in **less developed countries.** Increasing the density of a population makes it more difficult to find or provide food, shelter, services, and jobs. Increased density and industrialization are usually accompanied by increased levels of **air pollution** and **water pollution** and a general decrease in environmental quality. (See *Urban sprawl, Ribbon sprawl, Mass transit, Automobile fuel alternatives*)

Valdez Principles Refers to a set of environmental codes developed in 1989 by the Coalition for Environmentally Responsible Economies (CERES). It was dubbed the Valdez Principles after the **Exxon Valdez** oil spill disaster. Companies that sign this document agree to find ways to minimize pollutants, use renewable resources, and reduce the health risks to individuals and communities. Shareholders or boards of directors are asked to support these principles and urge their company to sign the document. Numerous companies have signed on, but no Fortune 500 companies to date. Write to 711 Atlantic Avenue, Boston, MA 02111.

vector An organism, such as a mosquito, that carries or transmits a pathogen to another organism (causing illness).

vegetarian A human *herbivore*; a person who does not eat meat. (See *Food chain*)

virgin habitat Refers to habitats not affected by human intervention, such as a virgin forest. (See *Wilderness*)

viruses All viruses are *parasites* that live inside another organism's cells. They reproduce by forcing the cell they live in (the host) to manufacture replicates of the original virus. Many viruses cause disease in organisms, including humans.

visible light spectrum That part of solar radiation that is visible to the human eye (between 380 and 780 nanometers).

visual pollution Visual pollution is one form of *aesthetic pollution,* which are highly subjective forms of pollution. Almost everyone would agree that an open landfill or garbage strewn on a beach is visual pollution. Most would say that billboards lining an otherwise beautiful countryside are visual pollution, but not the people advertising their products or services on the boards. Local and state governments often regulate what is and is not a form of visual pollution, hopefully based upon the feelings of the constituents who must live with the sights.

vitrification Also called glassification, refers to a technology that vitrifies (turn to glass) with heat and fusion, radioactive wastes. To be completed by 1999, the Hanford Waste Vitrification Plant will attempt to dispose of a large portion of the nuclear waste materials located at the **Hanford Nuclear Reservation.** (See *Nuclear waste disposal, Radioactive waste disposal*)

viviparous Refers to animals that produce live young from the body of the maternal parent. Humans are viviparous. (See *Oviparous, Ovoviviparous*)

volatile organic compounds (VOC) VOC is the collective name given to pollutants that are gases at room temperature. Many VOCs are emitted from common household products. All these gases contain carbon as the primary element. VOC's include **formaldehyde,** which is released from numerous building materials and products such as cosmetics. They also include benzene, xylene, and others, which are released from solvents (liquids that dissolve substances) such as household cleaning fluids, paint removers, and leather finishes. Electrical equipment and some plastics emit polychlorinated biphenyls, commonly called PCBs. If improperly disposed, these substances release the PCBs, which can either enter water supplies or become airborne and enter the body, where they accumulate in the fatty tissues.

Many VOCs cause symptoms similar to a common cold, including many respiratory ailments. Other such PCBs are believed to be carcinogenic and a serious threat. Many of the VOCs have not been thoroughly tested and therefore have little known about their inherent risks.

vulnerable species See *Threatened species.*

Waste Isolation Pilot Program (WIPP) Carlsbad, New Mexico, has been designated by Congress to become the first location for a permanent **nuclear waste disposal** site. It consists of a network of caverns carved into the geological formations of salt found deep beneath the surface. If it opens on schedule in 1997, it will begin accepting nuclear wastes that have been piling up at temporary sites around the country. The site is designed to accept "transuranic" nuclear wastes, which are solid materials such as tools, instruments, and building materials that are radioactively contaminated. Most of these wastes come from nuclear weapons production facilities around the country. The cost for the project is estimated to be about $1 billion. **NIMBY syndrome** activists might prevent this facility from opening in the near future, if at all. (See *Radioactive wastes, Plutonium pits*)

waste minimization Refers to reducing the volume of pollutants prior to discharging them into the environment. Instead of worrying about cleaning up wastes after they have been created ("end-of-pipe" technologies), waste minimization refers to ways of reducing the need to use raw materials in the first place. For example, chemicals used during manufacturing processes can be collected and re-used. This results in less **hazardous waste** to be disposed of later. (See *Source reduction, Disposal fees, Bottle bills*)

waste stream Refers to the flow of waste from its inception of becoming waste, such as being placed into a garbage container, food disposal, or on the floor of a manufacturing plant, to its final resting place in a **landfill** or some form of **incineration.** The largest component in our **municipal solid waste** stream is paper, which accounts for about 40 percent of the total and about 70 percent of the waste stream that ends up in landfills.

waste-to-energy power plants Power plants that burn waste products such as **municipal solid waste** or animal waste from feedlots to create electricity are called waste-to-energy power plants. Using solid waste helps reduce the need for **landfill** space and converts the energy available within the waste into valuable electricity. State of the art waste-to-energy **incineration**

facilities use intense heat to burn the waste thoroughly and have sophisticated air pollution devices to minimize the contaminants released in the emissions. These plants currently consume about 8 percent of the U.S. solid waste. Japan has built over 300 waste-to-burn plants that consume about 40 percent of its solid waste. (See *Biomass energy, Renewable energy*)

water Water is our planet's most abundant and important resource. Estimates place the total amount of water in all forms in the **biosphere** at 360 billion billion gallons; it is no surprise that the earth is called the water planet. Over 97 percent of all water is found in the oceans and commonly called salt water. The remaining 3 percent is fresh water. About 2 percent is in the ice sheets, .5 percent is in **groundwater**; only .02 percent is in surface waters (streams, rivers, ponds, lakes, reservoirs); .01 percent is in the soil; and even less is found in the atmosphere. Water covers over 70 percent of the earth's surface and is a major force in controlling the climate by storing vast quantities of heat. It is essential to all forms of life, which are composed primarily of water.

Water passes through the hydrologic or **water cycle**, which constantly replenishes the freshwater supply that most life depends upon. In spite of the vast quantity of all water, fresh water is in short supply in much of the world, including many parts of the U.S. where it is used heavily for **irrigation.**

The water available for human consumption (fresh water) is divided into **groundwater** and **surface water.** Much of the surface water is already polluted, and many underground **aquifers** are becoming contaminated, making the waters unfit for human consumption. Groundwater aquifers are being depleted, primarily due to irrigation in a process called **water mining.**

water cycle Water passes through a **biogeochemical cycle** as it passes through the **biosphere.** The energy from the sun drives this cycle by causing water from the earth's surface to enter the atmosphere; it does this by evaporation from bodies of water and the moisture in the soil and from transpiration, in which moisture from plants passes into the air. Warm air carries the moisture until the air cools, causing the moisture to form water droplets that return to the earth as precipitation. The water may fall back into bodies of water, be absorbed by the soil, where it is reabsorbed by plants, or percolate into the **groundwater.** If the soil cannot absorb the water, it runs off into streams and rivers and finally reaches the ocean.

The water cycle acts like a filtration system, purifying and removing salts from the water, producing the 3 percent of all water that is fresh and available for organisms to consume. (See *Aquifers*)

water diversion   Water diversion refers to physically diverting the flow of water from an area with an abundance of water to an area in short supply of water. Many of these projects are successful and pose little environmental threat compared with their rewards. However, since water is scarce in many agricultural regions, legal battles are fought over whose water gets diverted elsewhere. For example, Northern and Southern California have feuded over who gets the water that flows through the state's many aqueduct systems. Water diversion projects have also resulted in environmental disaster such as the old Soviet Union's **Aral Sea.**

water mining   Water mining pertains to **groundwater** that is pumped out of the earth for human use faster than natural processes can replace it, lowering the level of the **water table.** In some areas of the San Joaquin Valley in California, the water table has dropped over 300 feet and the land itself is dropping a few fractions of an inch each year. It is estimated that water mining is occurring in 35 of the 48 contiguous states in the U.S. This depletion occurs due to our water demands primarily for agricultural **irrigation,** but also for industry and residential purposes. (See *Water use, Aquifers, Groundwater*)

water pollution   Water pollution occurs when the natural quality of water is degraded in some way. This degradation results in damage or the destruction of **aquatic ecosystems** and/or makes the water resource unfit for human consumption.

Water pollution is caused by chemicals and substances added to the water such as **heavy metals** and acids from **industrial waste** that are dumped into bodies of water. Nitrates and phosphates from **fertilizers** and toxic compounds from **pesticides** run off or float into bodies of water, contaminating it. Many **oil**-based products such as gasoline and **plastic** products that make their way into bodies of water also cause pollution.

Water pollution may also be caused by organic wastes, which include food and human waste found in **sewage** that makes its way into water. This material harbors and encourages the growth of disease-causing microbes. Organic waste often dramatically increases the abundance of bacteria that use up all the oxygen available in the water.

Another cause of water pollution is radioactive wastes. When radioactive materials are produced, used, or disposed of, they can enter bodies of water, killing aquatic organisms or causing long-term effects. Water used to cool manufacturing processes and power plants causes thermal water pollution. It becomes heated and damages ecosystems by changing the normal temperatures and lowering the oxygen levels.

Finally, excessive **sediment** due to manmade **soil erosion** damages or destroys aquatic ecosystems and makes it less suitable for human consumption.

water table Refers to the upper portion of the **zone of saturation,** which is readily accessible for human consumption. (See *Aquifers*)

water treatment Refers to the process of making water suitable for human consumption. Several methods are used, such as sedimentation (when solids settle out), coagulation (when masses called floc accumulate and are separated out), filtration (when contaminants are filtered out by passing the water through sand or similar material), or disinfection (when a chemical such as chlorine is used to kill **pathogens**). (See *Chlorination, Sewage treatment*)

water use for human consumption Water is used for human consumption in four ways: 1) **domestic water** use; 2) agricultural **irrigation**; 3) **industrial water use**; and 4) **in-stream water use.** About 70 percent of the water used worldwide (excluding in-stream use) is for irrigation. The U.S. uses the most water of any nation, followed by China and India.

waterlogging In areas where layers of soil are exceptionally impermeable and water cannot drain properly, the soil becomes waterlogged, retaining excess quantities of water.

watershed The watershed (also called catchment area or drainage basin) refers to the land surrounding a lake or river. This area is responsible for most of the water entering the lake or river. Precipitation falls on the watershed, which in turn delivers the water to the lake or river as runoff. The region's size, shape, and vegetation are responsible for the amount and type of water to enter the body. For example, watersheds in a forested region will deliver water that is rich in organic matter, but only after the trees have soaked up their share first. Bare rocks on a mountainside will rapidly deliver water with few organic nutrients as the water rushes through the watershed into the water. (See *Aquatic ecosystems, Biomes*)

watt Electric power is measured in units called watts. A watt is equal to one *joule* per second. The total generating capacity of a power plant is measured in kilowatts (kW) for 1,000 watts, and megawatts (MW) for one million watts. For example, the Grand Coulee Dam, which produces electricity by *hydroelectric power,* creates over 6,100 MW of power as compared to some small, wood-burning power plants (*biomass power*) that produce about 250 MW of power.

The ongoing quantity of electricity produced is measured in kilowatt hours (kWh) and is often used to compare the costs of various electric production methods. For example, *nuclear power* plants produce electricity at about 12.5 cents per kWh, wood power plants at about six cents per kWh, and *photovoltaic cells* at over 28 cents per kWh.

wave power A few small experimental wave power plants have been built that capture the energy from the motion of waves. Although limited to coastal regions, this form of power shows promise for further development in coastal regions. (See *Moon power, Tide power, Hydroelectric power, Alternative energy*)

weathering Refers to the physical, chemical, or biological wearing of rock contributing to *soil* formation. (See *Parent rock, Soil particles, Phosphorus cycle*)

weatherstripping In most older homes and some new homes, cracks, holes, and spaces exist in the structure that allow warm air to escape in the winter and cool air to escape in summer. These air leaks account for about 35 percent of the total lost energy from a typical house. The easiest and most cost-effective method to minimize this lost energy is with weatherstripping. Weatherstripping consists of narrow bands of metal, vinyl, rubber, felt, or foam that are applied to the openings. It is usually applied to joints between various structures in the house where the leaks occur, such as window and door joints, wall joints near the foundation, joints around a fireplace, attic openings, or around electrical outlets and service boxes. (See *Thermal insulation, Insulation, Super-insulation*)

weed Any plant growing where humans don't want it to grow. (See *Herbicides, Pesticides*)

wetlands Wetland is a collective term for a number of **habitats,** including marshes, swamps, and bogs. Wetlands are divided into two types: **inland wetlands** and **coastal wetlands.** At least one of the two can be found in all 50 states. These habitats contain some of the most productive and useful ecosystems found anywhere. Wetlands are to nature what farms are to man. They produce huge volumes of food in the form of plants, both alive and—more importantly—dead and decomposing as **detritus.** This **biomass** acts as the first level of a **food chain** that feeds everything from small invertebrates to large predatory fish such as striped bass.

Acting like filters, wetlands maintain the quality of water in rivers and streams by removing and retaining nutrients, and helping to decompose waste. They also reduce the amount of sediment that enters a body of water.

Wetlands minimize the effect of flooding by storing huge volumes of excess water, which protects all forms of wildlife downstream. They also buffer the shoreline from erosion and protect entire coastal areas from washing away.

Unfortunately, wetlands were viewed as wasteland, of little value, as late as the early 1970s. Because of this misconception, vast quantities of our wetlands were destroyed within the last few decades. Less than half the number of wetland acres that existed in the 1600s exist today. From the mid 1950s to the mid 1970s, over 11 million acres of wetlands were destroyed, with most drained for agricultural use. The remaining wetlands are now protected under sections of the **Clean Water Act** but have been under attack once again in recent years by proponents of development.

The **biodiversity** of wetlands is among the most abundant in the world, similar in **net primary production** with tropical rain forests. Many species are unique to these habitats. Inland wetlands are inhabited by numerous species of freshwater fish and wildlife. Ducks, geese, and many songbirds feed, nest, and raise their young in wetlands, and most recreational fish spawn in wetlands. Many mammals also feed and live in inland wetlands.

Coastal wetlands are habitats for estuarine and marine fish, shellfish, waterfowl, and many birds and mammals. Most commercial and game fish use coastal marshes and estuaries for spawning. (See *Estuary and coastal wetlands destruction*)

wilderness The U.S. Congress defined wilderness in the Wilderness Act of 1964 as "an area where the earth and its community of life are untrampled by man, where man himself is a visitor and does not remain". (See *Wilderness, top ten countries for; National Park and Wilderness Preservation System*)

Wilderness Society, The The Wilderness Society's purpose has remained the same since it was founded by the famous conservationist and writer *Aldo Leopold,* who instilled the belief that land is not a commodity but a valuable resource. Founded in 1935, the society has 400,000 members and spends much of its efforts working with Congress to preserve and manage federally owned lands. They are also involved in conservation education. The annual membership fee is 30 dollars. Write to 900 17th Street NW, Washington DC 20006.

wilderness, top ten countries for Those countries (and land masses) with the largest amount of remaining **wilderness** are as follows: (in square miles/percent of total land) 1) Antarctica (5.2 million/100%); 2) nations that comprised the Soviet Union (3 million/33.6%); 3) Canada (2.5 million/64.6%); 4) Australia (917,000/29.9%); 5) Greenland (869,000/99.9%); 6) China (843,000/22%); 7) Brazil (808,000/23.7%); 8) Algeria (561,000/59%); 9) Sudan (317,000/31.7%); and 10) Mauritania (285,000/69.2%). The U.S. comes in 16th (176,200/4.7%).

wildlife Refers to all the uncultivated vegetation and nondomesticated animals in an area.

Wildlife Habitat Enhancement Council (WHEC) The WHEC, established in 1988, is a nonprofit, nonlobbying organization that aids corporations to improve their lands for wildlife, since they own as much as a third of all private land in the United States. The council has designed a program that links company managers with environmental experts to create wildlife projects. This involvement helps wildlife thrive and enhances the company's environmental image and establishes better local community relations. Write to WHEC, 1010 Wayne Avenue, Suite 1240, Silver Spring, MD 20910.

wildlife management Refers to maintaining or promoting the survival of wildlife. Management may pertain to one species or to all the organisms in an area. All the wildlife may be managed with the hope of maintaining natural populations against human intervention. Certain animals might be managed for a specific purpose such as hunting. Managing a particular animal means studying and understanding their food and water requirements, and forms of protection or cover from the elements, and from *predators.* It also means understanding the *population dynamics* of the area.

Once these studies are complete, decisions can be made about how best to manage the wildlife. The methods used can range from simply leaving the natural populations alone, to introducing additional numbers of some species, or reducing the numbers of some species.

Managing wildlife often includes habitat management. This may involve building structures that provide cover from predators, or the destruction of some part of a habitat to encourage or discourage a certain population (for example, cutting mature trees to allow new saplings to grow, providing food for deer). Habitat management of some animals, such as migratory birds, has required the establishment of international agreements since many of these birds can have habitats in three countries each year as they migrate along their *flyways.* (See *Forest Service, Forest management, integrated; Animal Damage Control Act*)

wind power Wind is a *renewable energy* resource that has been used for hundreds of years, both in the U.S. and worldwide. Wind energy is air- and water-pollution free. It is one of the most energy-efficient technologies, converting over 90 percent of the available energy into useable energy. It is already cost-competitive with other sources of energy at 7 to 11 cents per *kwh,* making it the least expensive of all the *renewable energy* sources.

Very little wind energy is produced in the U.S. today. Most of the existing production occurs in California, where over 1,300 *MWs* are produced by 15,000 wind turbines. This supplies about 1 percent of the state's total needs, which is enough to power all of San Francisco. Although wind could theoretically supply all U.S. energy needs, experts believe it can realistically supply about 20 percent of our total energy within the next 20 years.

Wind turbines located at high-wind locations, with an average wind of 15 miles per hour, are already highly efficient. However, most U.S. sites are considered moderate sites with winds averaging between 12 and 15 miles per hour, or low-wind sites with averages below 12 miles per hour. Technological advances are needed to make these low and moderate sites more economically feasible if wind power is to flourish in the U.S.

Most wind turbines consists of two or three blades, a rotor, a transmission, an electric generator, and controls, mounted on a tower that is 15 to 30 meters (50 to 100 feet) high. The rotor is capable of turning, to align itself with the wind. Most newer blades are made of fiberglass while older models were made of aluminum. The larger the turbine, the more efficient the device. This makes them more practical for large utility companies than for small power plants or individual use.

Environmental problems are nil with the only complaint being some **noise pollution.** (See *Wave power, Moon power, Tide power*)

windbreaks   When **soil erosion** is caused by wind instead of water, windbreaks are used as a **soil conservation** technique. Windbreaks, also called shelterbelts, are long rows of trees that break the wind. They are usually used in wide open, flat lands such as the Great Plains, where wind often causes erosion. (See *Dust bowl*)

World Resources Institute (WRI)   WRI is an independent research and policy institute established in 1982 to aid governments, environmental organizations, and private businesses to better understand how society can meet basic human needs and nurture economic growth without depleting the earth's natural resources. WRI presently addresses six broad areas: climate, energy, and pollution; forests and biodiversity; economics; technology; resource and environmental information; and institutions that augment policy recommendations for groups working in natural resource management. Write to 1709 New York Avenue NW, Washington DC 20006.

Worldwatch Institute   Worldwatch Institute is a world leader in providing global environmental information to decision-makers and the public. Its analysis of global management issues and the advice it renders is respected by world leaders and understandable to the public. Its primary focus is on issues regarding population, energy, food policy, and environmental quality. Its annual *State of the World* publication has become the standard reference for environmental policy decision-making. The institute was founded in 1975 and its annual membership fee is 25 dollars. Write to 1776 Massachusetts Avenue NW, Washington DC 20036.

xenobiotic A foreign organic substance, such as an organic pesticide, found in **drinking water**.

Xerces Society This organization is dedicated to preventing the extinction of invertebrates, including many **insects,** caused by human intervention. The Xerces Society has three programs: conservation science, education, and public policy. Write to 10 Southwest Ash Street, Portland, OR 97204. (See *Insects*)

xerophyte A plant adapted for life in a **desert.**

yellowcake Low-grade uranium ore is mined to produce fuel for **nuclear reactors.** The ore goes through a milling process, which involves crushing and treating it with solvents to concentrate the uranium. The resulting mixture is called yellowcake, which then goes through an enrichment process. (See *Nuclear fuel cycle, Tailings, uranium mine*)

YIMBY—FAP Acronym for Yes, In My Backyard—For a Price. Waste disposal firms have found that they can often overcome the **NIMBY syndrome** by offering large financial inducements to towns that are then willing to allow large **landfills** or **incineration** plants.

Yosemite National Park Yosemite became the third national park in 1890 (following Yellowstone and Sequoia) and includes about 750,000 acres. It is home to two of the world's highest waterfalls, has giant sequoias, and mountain peaks reaching over 13,000 feet. Wildlife includes mountain lions, black bear, and about 200 species of birds. It also blossoms with over 1,300 species of flowering plants. Ninety-four percent of the park is designated as wilderness and therefore is off-limits to road building or other development. About 3.5 million people visit Yosemite each year.

Yucca Mountain The Yucca Mountains of Nevada have been designated by Congress as a permanent **nuclear waste disposal** site for high-level nuclear waste. It is scheduled to open in the year 2010 but is meeting with considerable opposition. (See *Radioactive waste disposal, Waste Isolation Pilot Program, NIMBY syndrome*)

zero discharge   Refers to the complete elimination of pollutants from entering the environment. Zero-discharge activists believe that all production processes and agricultural methods should be reformulated so that absolutely no hazardous substances are discharged into the air, soil, or water. (See *Pollution*)

zero population growth   Refers to a population with a growth rate equal to zero, which is accomplished by balancing births (plus *immigration*) and deaths (plus *emigration*). The current worldwide people population is about 5.4 billion, and we will be adding close to 100 million people to our planet each year. In roughly 40 years, the world population will be about 10 billion, if current trends hold true.

   ZPG has been reached in a few countries, such as Italy, Germany, and Japan, where the fertility rate has been below the replacement level of 2. The U.S. has a fertility rate of 2.1. The world rate is still a staggering 3.9, even though it has dropped considerably over the past 20 years. Many Third World countries such as Ethiopia and Tanzania have fertility rates well above 6. (See *Doubling-time in human populations, Population growth*)

Zero Population Growth (ZPG)   ZPG is an organization that advocates stabilizing the population in the U.S. and the world. ZPG publishes several reports and teaching kits that describe how to stop population growth and explain its overall importance to the environment. Write to 1400 16th Street NW, Suite 320, Washington DC 20036.

zone of saturation   Refers to that portion of the earth's crust that is saturated with water, called *groundwater.* The zone of saturation within the earth contains about 40 times as much water as all the *surface water* (ponds, lakes, streams, and rivers). (See *Water, Aquifers*)

zones of life   Dating back to the turn of the century, scientists have tried to define the life that exists in different regions of the planet using a variety of methods. Different names have been given to these various zoning methods.

"Formations" specifically refer to the plant life that exists in major regions and include deserts and coniferous forests. "Realms" refer to the distribution of animals in regions of the world, and include areas such as nearctic and neotropical.

The original "life zones" system, which was described around the turn of the century, attempts to combine both the plants and animals into one scheme and divides the earth east and west into transcontinental bands. A more recent theory is the Holdridge Life Zone System, which is a complex system taking into account numerous variables and includes altitude as well as latitude.

Probably the most accepted term for describing zones of life is the *biome.* Biomes use plant formations (as described above) and the animal life associated with these plants to describe distinct zones of life on the planet.

zoning, land-use   The specific use of land is often regulated by local or regional zoning laws. Land can be zoned, for example, as agricultural, commercial, industrial, recreational, or residential. The individuals who decide on the zoning regulations should, in theory, be qualified professional planners with access to environmental consultants. Those individuals responsible for zoning regulations should be individuals who are responsible for the economic well-being of the area and who are concerned and knowledgeable about the area's environmental well-being. (See *Urban open space, Urban sprawl, Ribbon sprawl*)

zoo   Zoos were originally built to satisfy human curiosities about wild animals. Since this was their primary purpose, most zoos were virtual prisons to their inhabitants. With few people opposing this type of facility, zoos remained in this state for decades. During the 1960s, animal rights organizations protested how zoos treated their prisoners. The public began to listen and so did the zoos. Instead of being treated as curiosities, animals were viewed as part of a habitat, and treated as the most fascinating aspect of that habitat.

In most large zoos, animals are not confined to cages and are treated humanely. Although it is still a debatable issue, many people believe the educational advantages and the heightened public awareness raised by zoos make them worthwhile.

In addition and possibly more important, zoos have become a major force in protecting **endangered** and **threatened species** from extinction with breeding and conservation programs. For example, the **Smithsonian Institution,** through the National Zoo, has successfully reintroduced the golden lion tamarin back into the Brazilian rain forests. Many other zoos are playing important roles in saving other species.

Zoo Doo   Zoo Doo is a unique and entrepreneurial alternative to chemical *fertilizers* for your houseplants. Economically, it's not a great investment since it costs about twice as much as other fertilizers, but it is a good environmental investment and makes an interesting gift. Offered by a company in Memphis, Tennessee, it is a mixture of elephant and rhino manure, collected in a number of zoos around the coun-  try. The manure is decomposed into an organic *compost* and packaged for sale. The Woodland Park Zoo in Seattle converts a portion of the two tons of manure produced each day by its residents into Zoo Doo, which saves the zoo over $30,000 in landfill fees each year and generates revenue as well. Call (800) I LUV DOO. (See *Kricket Krap, Composting, large-scale*)

zoogeography   Pertains to the study of where animals live and, furthermore, why they live where they do; a subdivision of biogeography.

zoology   The study of animals.

zoonosis   A disease that can naturally be transmitted from animals to man.

zooplankton   Refers to the microscopic animals found in *plankton,* which consists of crustaceans, rotifers, and protozoans. These animals usually fill the role of primary consumers in many *aquatic ecosystems.* (See *Phytoplankton*)

A SPECIAL NOTE TO EDUCATORS

This book can be used for a high school or college (non–science major) course. It can supplement an existing biology text when additional emphasis is desired in the environmental sciences, or it can be used on its own, to customize an environmental-science course. Students looking for environmental term papers or research topics could use this book to select a topic and begin their work.

To assist you in developing a course, a few topic outlines are suggested below. They can be followed as is, or modified to suit your needs. Outlines are interdisciplinary, with entries about the sciences, the people, and the laws involved in the "big picture." This is the only way to truly understand an environmental issue.

Only ten outlines are listed here. The number of potential topic outlines is as vast as your imagination. I am interested in hearing about any outlines you've created and used for class.

(Every item, at all levels, is an entry in the book.)

TOPIC OUTLINE: PESTICIDES

1. Pesticides
2. Pesticide dangers
        Bioaccumulation
        Biological amplification
        Pesticide regulations
        Pesticide residues on food
        Circle of poison
        Nontarget organisms
        Carson, Rachel
        DDT
        Public Voice for Food and Health Policy
3. Integrated pest management
        Insect sterilization
        Sex attractants
        Resistant crops
        Natural insecticides

Biological control
Biological control methodologies
4. Insecticides
Chlorinated hydrocarbons
DDT
Organophosphates
Carbamates
Pyrethroids
5. Herbicides
Hormone weed killers
Agent Orange
6. Fungicide
7. Rodenticide

TOPIC OUTLINE: RECYCLING

1. Recycling
2. Bottle bills
3. Grass cycling
4. Paper recycling
5. Plastic recycling
Styropeanuts
Aseptic containers
6. U.S. PIRG
7. Tires, recycled
8. Motor oil recycling
9. Automobile recycling
10. Appliance recycling
11. Copier toner cartridges, recycled
12. Composting, large-scale
13. Telephone pole disposal
14. *Garbage* magazine
15. Kricket Krap
16. Zoo Doo

TOPIC OUTLINE: CITY LIFE

1. Urbanization and urban growth
2. Urban sprawl
Tract development
Ribbon sprawl

3. Megalopolis
4. Farmland lost
5. Mass transit
6. Automobile fuel alternatives
7. Urban heat island
8. Zoning, land-use
9. Alliance for a Paving Moratorium
10. Ecocity
    Urban open spaces
    Greenways
    Rails-to-Trails Conservancy
11. Green cities
    Nature centers
    Touchpools, aquarium
    Zoo

Topic Outline: Food, Farming, and Health

1. Monoculture
2. Fertilizers
3. Pesticides
4. Pesticide dangers
5. Irrigation
6. Drip irrigation
7. Sustainable agriculture
8. Integrated pest management
9. Organic farming
    Crop rotation
    Strip farming
    Conservation tillage
    Terracing
10. Organic fertilizer
    Green manure
    Night soil
    Compost
11. Organic food supermarkets
    Organic labelling
    Hydroponic aquaculture
12. Food irradiation
13. Factory farms
14. Organic beef
15. Center for Science in the Public Interest

Topic Outline: In the Home and Office

1. Indoor pollution
2. HVAC systems
3. Outgassing
4. Humidifier fever
5. Laser printer ozone
6. Formaldehyde
7. Asbestos
8. Electromagnetic radiation
9. Radon
10. Baubiologie
11. Healthy homes
12. Environmental tobacco smoke
13. HEAL
14. Drinking water
15. Domestic water conservation
    Refrigerator door
    Thermal insulation, building
    Insulation
    Caulking
    Weatherstripping
    R-value
    Super-insulation

Topic Outline: Water

1. Water
2. Water use for human consumption
3. Water pollution
4. Clean Water Act
5. Coastal Society, The
6. Aquatic ecosystems
7. Wetlands
    Inland wetlands
    Estuary and coastal wetlands destruction
    Everglades
    Douglas, Marjory Stoneman
8. Oceans
    Cousteau, Jacques-Yves
    Ocean zone ecosystems
    Ocean pollution
    Center for Marine Conservation

9. Drinking water
  Surface water
  Groundwater
  Groundwater pollution
  Aquifers
10. Water treatment
11. Chlorination
12. THM

Topic Outline: Waste Disposal

1. Garbage
2. Municipal solid waste
3. Municipal solid waste disposal
  Disposal fees
  Green tax
4. Landfills
5. Landfill problems
6. Incineration
7. Waste-to-energy power plants
8. Incineration problems
  Bottom ash
  Fly ash
  Char
9. Sewage
  Sewage treatment
  Sludge
  Sludge disposal
  Effluent
10. Composting, large-scale
11. Waste minimization
12. Source reduction
13. Hazardous waste
  Love Canal
  Gibbs, Lois
14. INFORM
15. Waste Isolation Pilot Program

Topic Outline: The Environmental Movement

1. Environmental literacy
2. Environmentalist

3. The League of Conservation Voters
        Greenspeak
        Greenscam
4. NIMBY syndrome
5. NOPE
6. Environmental organizations
7. Ethics, environmental
8. Worldwatch institute
9. Izaak Walton League
10. Eco-conservative
11. Eco-terrorism
12. Deep ecology
13. Gaia
14. Philosophers, environmental
        Muir, John
        Leopold, Aldo
        Brown, Lester

TOPIC OUTLINE: RENEWABLE ENERGY

1. Energy sources, historical
2. Energy consumption, historical
3. Fossil fuels
4. Alternative energy
5. Renewable energy sources
6. Solar power
        Passive solar-heating systems
        Active solar-heating systems
        Solar thermal power
        Photovoltaic power
        Solar ponds
7. Wind power
8. Hydroelectric power
        Moon power
        Wave power
        Ocean thermal energy conversion
9. Geothermal power
10. Biomass energy
        Biomass direct combustion
                Waste-to-energy power plants
                Energy plantations

Biofuels
    Biochemical conversion of biofuels
      Biogas
      Methane digesters
    Thermochemical conversion of biofuels
      Gasification
      Syngas
  Plant-oil fuels
  Sunflower oil fuel

TOPIC OUTLINE: NUCLEAR POWER

1. Energy sources, historical
2. Energy consumption, historical
3. Fossil fuels
4. Alternative energy
5. Nuclear power
6. Nuclear reactors
7. Nuclear fission
    Light-water nuclear reactor
    Heavy water nuclear reactor
    High-temperature, gas-cooled nuclear reactor
    Breeder reactor
8. Nuclear waste disposal
    Waste Isolation Pilot Program
    Yucca Mountain
9. Nuclear reactor safety and problems
    Meltdown
    Chernobyl
    Three Mile Island
    Decommissioning nuclear reactors
    Tailings, uranium mining
    Union of Concerned Scientists
10. Nuclear fuel cycle
    Yellowcake
    Nuclear fuel cycle hazards
11. Nuclear fusion

SOURCE MATERIAL

Brower, Michael. *Cool Energy*. Cambridge, MA: Union of Concerned Scientists, 1990.

Brown, Lester R. *State of the World*. New York: W. W. Norton & Company, 1991.

——*State of the World*. New York: W. W. Norton & Company, 1992.

——*Saving the Planet*. New York: W. W. Norton & Company, 1991.

Buzzworm. *1992 Earth Journal*. Boulder, CO: Buzzworm Books, 1992.

Clapham, W. B. *Natural Ecosystems*. New York: Macmillan, 1973.

Coffel, Steve, and Karyn Feiden. *Indoor Pollution*. New York: Fawcett Columbine, 1990.

Enger, Eldon D. *Environmental Science*. Dubuque, IA: W. C. Brown Publishers, 1991.

Frost, S.W. *Insect Life*. New York: Dover Publications, Inc., 1959.

Grupenhoff, John T., and Betty Farley. *Congressional Directory: Environment*. Bethesda, MD: Grupenhoff Publications, Inc., 1989.

Henry, J. Glynn, and Gary W. Heinke. *Environmental Science and Engineering*. Englewood Cliffs, NJ: Prentice Hall, 1989.

Kraus, David. *Concepts in Modern Biology*. New York: Cambridge Book Company, 1974.

Lapedes, Daniel N. *Encyclopedia of the Geological Sciences*. New York: McGraw-Hill, 1978.

League of Conservation Voters. *Vote for the Earth*. Berkeley, CA: Earthworks Press, 1992.

Lean, Geoffrey, Don Hinrichsen, and Adam Markham. *Atlas of the Environment*. New York: Prentice-Hall Press, 1990.

Lewis, Walter H. *Ecology Field Glossary, A Naturalist Vocabulary*. Westport, CT: Greenwood Press, 1977.

Lincoln, R. J., and G. A. Boxshall. *The Cambridge Illustrated Dictionary of Natural History*. Cambridge, MA: Cambridge University Press, 1990.

Miller, G. Tyler, Jr. *Environmental Science, Sustaining the Earth*. Belmont, CA: Wadsworth Publishing Company, 1991.

National Audubon Society. *Audubon Wildlife Report 1987.* Orlando, FL: Academic Press Inc., 1987.

National Wildlife Federation. *1990 Conservation Directory.* Washington DC, 1990.

National Wildlife Federation. *1993 Conservation Directory.* Washington DC, 1993.

National Academy of Sciences. *One Earth, One Future.* Washington DC: National Academy Press, 1990.

Null, Gary. *Clearer, Cleaner, Safer, Greener.* New York: Villard Books, 1990.

Parker, Sybil P. *Encyclopedia of Science and Technology.* New York: McGraw-Hill, 1992.

Pelczar, Michael J., Jr., and Roger D. Reid. *Microbiology.* New York: McGraw-Hill, 1965.

Raven, Peter H., Ray F. Evert, and Helena Curtis. *Biology of Plants.* New York: Worth Publishers, Inc., 1976.

Rifkin, Jeremy. *Green Lifestyle Handbook.* New York: Henry Holt & Company, 1990.

Rittner, Don. *Ecolinking.* Berkeley, CA: Peachpit Press, 1992.

Rogers, Adam. *The Earth Summit: A Planetary Reckoning.* Los Angeles: Global View Press, 1993.

Seredich, John. *Your Resource Guide to Environmental Organizations.* Irvine, CA: Smiling Dolphins Press, 1991.

Smith, Robert Leo. *Elements of Ecology.* New York: HarperCollins Publishers Inc., 1992.

Stein, Edith C. *The Environmental Sourcebook.* New York: Lyons & Burford, 1992.

Villee, Claude A. *Biology.* Philadelphia: W. B. Saunders Company, 1977.

Weisberger, Berbard A. *Family Encyclopedia of American History.* Pleasantville, NY: Reader's Digest Association, 1975.

World Resources Institute. *World Resources, A Guide to the Global Environment.* New York: Oxford University Press, 1992.

Worldwatch Resources Institute. *Environmental Almanac.* New York: Houghton Mifflin Company, 1992. Also 1993.

Wright, John W. *The Universal Almanac.* Kansas City, MO: A Universe Press Syndicate Company, 1992.

Journals & Reports

Driesche, Roy Van and Eileen Carey. *Opportunities for Increased Use of Biological Control in Massachusetts.* Amherst, MA: University of Massachusetts, 1987.

Gray, Irving, Joseph Neale, and Lisa Gray. *Journal of the Washington Academy of Sciences*. Arlington, VA: Washington Academy of Sciences, 1987.

The following magazines were used extensively:

Audubon. Boulder, CO: National Audubon Society.
BioCycle. Emmaus, PA: J.G. Press, Inc.
Buzzworm. Boulder, CO: Buzzworm Inc.
Earth. Waukesha, WI: Kalmbach Publishing Co.
E Magazine. Westport, CT: Earth Action Network, Inc.
Environment. Washington, DC: Heldreff Publications.
Garbage. Gloucester, MA: Old House Journal Corp.
In Business. Emmaus, PA: J.G. Press, Inc.
Issues in Science and Technology. Washington DC: National Academy of Sciences.
National Parks. Washington DC: National Parks and Conservation Assoc.
Newsweek. New York: Newsweek, Inc.
Popular Science. New York: Times Mirror Magazines, Inc.
Sierra. San Francisco: Sierra Club.
Smithsonian. Washington DC: Smithsonian Institution.
U.S. News & World Report. Washington DC: U.S. News & World Report, Inc.
Wilderness. Washington DC: Wilderness Society.
Wildlife Conservation. Bronx, NY: New York Zoological Society.
Worldwatch. Washington DC: Worldwatch Institute.

The *New York Times* was used extensively to update entries.

LIST OF ENTRIES

ABOUT THE AUTHOR

H. STEVEN DASHEFSKY is an adjunct professor of environmental science at Marymount College in Tarrytown, New York, and a former employee of the Environmental Protection Agency in Washington, D.C. He is the founder of the Center for Environmental Literacy, which was created to inform and educate the general public and business community about today's complex environmental issues. He holds degrees in biology and entomology. Steve lives in Ridgefield, Connecticut, with his wife and two children.

THE CENTER FOR ENVIRONMENTAL LITERACY

Using a variety of media, the Center for Environmental Literacy offers informative and entertaining material about the environment: newsletters are available for the general public, corporations, and schoolteachers; brochures have been designed to inform a business's customers or employees about specific environmental topics; H. Steven Dashefsky conducts seminars and in-house workshops on environmental literacy. For more information, contact the Center for Environmental Literacy, 383 Main Street, Ridgefield, Connecticut 06877, telephone (203) 438-8080.

 ENVIRONMENTAL LITERACY